Institute of
Terrestrial
Ecology

Natural Environment Research Council

Biological recording of changes in British wildlife

ITE symposium no. 26

Proceedings of a Conference held on 13 March 1990 to celebrate the 25th Anniversaries of the Biological Records Centre and the Natural Environment Research Council

Edited by

Paul T Harding

London: HMSO

The INSTITUTE OF TERRESTRIAL ECOLOGY (ITE) is one of 15 component and grant-aided research organizations within the NATURAL ENVIRONMENT RESEARCH COUNCIL. The Institute is part of the Terrestrial and Freshwater Sciences Directorate, and was established in 1973 by the merger of the research stations of the Nature Conservancy with the Institute of Tree Biology. It has been at the forefront of ecological research ever since. The six research stations of the Institute provide a ready access to sites and to environmental and ecological problems in any part of Britain. In addition to the broad environmental knowledge and experience expected of the modern ecologist, each station has a range of special expertise and facilities. Thus, the Institute is able to provide unparallelled opportunities for long-term, multidisciplinary studies of complex environmental and ecological problems.

ITE undertakes specialist ecological research on subjects ranging from micro-organisms to trees and mammals, from coastal habitats to uplands, from derelict land to air pollution. Understanding the ecology of different species of natural and man-made communities plays an increasigly important role in areas such as improving productivity in forestry, rehabilitating disturbed sites, monitoring the effects of pollution, managing and conserving wildlife, and controlling pests.

The Institute's research is financed by the UK Government through the science budget, and by private and public sector customers who commission or sponsor specific research programmes. ITE's expertise is also widely used by international organizations in overseas collaborative projects.

The results of ITE research are available to those responsible for the protection, management and wise use of our natural resources, being published in a wide range of scientific journals, and in an ITE series of publications. The Annual Report contains more general information.

P T Harding
Institute of Terrestrial Ecology
Monks Wood Experimental Station
Abbots Ripton
Huntingdon
Cambs PE17 2LS

Tel: 048 73 381

Contents

Foreword

One of the many events marking the 25th Anniversary of the Natural Environment Research Council in 1990 was an enjoyable and informative Conference hosted by the Linnean Society at their splendid rooms in Burlington House in London. The Conference was organised by Paul Harding, Head of the Institute of Terrestrial Ecology's (ITE) Biological Records Centre, also in celebration of the 25th Anniversary of the Centre.

It was my particular pleasure to welcome participants to the lunch-time conversazione of that Conference, and it gives me equal pleasure to add a Foreword to this volume of proceedings.

The Natural Environment Research Council (NERC), as our name implies, is in the business of research rather than data collection or monitoring for its own sake. That being said, much of the original research we pursue focuses on changes in characteristics of the natural environment, both physical and biological. Many of these changes are almost imperceptible and their study relies on careful and accurate observation, measurement, or sample collection over long periods of time. This is true for studies of climate and pollution as for changes in landscape and its effect on the indigenous plant and animal communities that make up the diversity of British wildlife.

The Biological Records Centre (BRC) at Monks Wood takes its rightful place alongside the other time-series databases and recording groups in NERC's extensive and varied armoury of environmental data, information and expertise. What makes BRC perhaps unique is its foundation, based upon the efforts of a veritable army of enthusiastic volunteers and natural history societies around the country, who observe and record their natural environment and sustain the 60 or so national recording schemes that cover the range of taxonomic groups. Coupling this network of observers with the expertise available to and within BRC, we have, therefore, an enviable 'community environment programme'.

The high-quality products of the BRC, be they the database itself, the *Atlas* series, reports, or other publications, all serve a dual purpose. They provide the foundation upon which research into environmental change is being based and they act as a testimony to, and a reference for, the scientists and lay naturalists whose contributions made their production possible.

The papers and summaries of posters in this volume cover the origins and history of BRC and its present context in ITE's recently formed Environmental Information Centre. The theme of biological recording of changes in British wildlife is particularly appropriate in these times of rapid changes in the natural environment. However, more than one paper highlights the fact that the flora and fauna of Britain are, and always have been, in a state of change. What is perhaps central to the dual Anniversary is that the last 25 years have seen unprecedented and rapid changes in land use and environmental management. The Biological Records Centre has a unique and continuing role in contributing to NERC's research on these changes through the involvement of the thousands of volunteer specialists who contribute to its database.

I anticipate that the next 25 years of BRC's work will be recognised as being of increasing importance in the study of terrestrial change and adaptation.

Professor John Knill
Chairman
Natural Environment Research Council

Preface

The Conference and conversazione, from which this publication originated, was a rewarding event to have conceived and organised, and in which to have participated. It was held at the very end of the Biological Records Centre's 25th Anniversary year, but this timing allowed the event to embrace the 25th Anniversary celebrations of the Centre's parent body, the Natural Environment Research Council.

Despite being organised at comparatively short notice, all the first-choice speakers and chairmen were able to participate, and a capacity audience filled the lecture room of the Linnean Society at Burlington House in London. The domestic arrangements at Burlington House were undertaken by John Marsden, Secretary of the Linnean Society, and his staff: I am most grateful for the welcome and hospitality shown to us and for their unobtrusive efficiency throughout the day.

Our chairmen of sessions were Professor Mike Roberts, Professor Sam Berry and Professor Mike Claridge, and I am very grateful to them for ensuring that the speakers followed the timetable and that the discussions were not too discursive. It was particularly pleasing that Mike Claridge, as President of the Linnean Society, was able to invite Professor John Knill to give an opening address at the conversazione.

It is my pleasure to be able to thank all those who presented papers at the meeting and who have contributed to this publication, those who have presented posters and have written up brief accounts, and those behind the scenes, particularly my colleagues in BRC and Mrs Joyce Rushton who has typed these proceedings.

At the end of the presentation, Frank Perring proposed that 'filial' greetings should be sent to the three people who had such crucial roles in the foundation of co-ordinated biological recording which is epitomised by BRC. The meeting was sorry that all three, for reasons of health, were unable to attend, but I was able to contact Professor Roy Clapham, Max Nicholson and Dr Max Walters, following the meeting, to convey our good wishes and thanks. Sadly, Roy Clapham died at the end of 1990.

P T Harding
Head
Biological Records Centre
Environmental Information Centre
Institute of Terrestrial Ecology
Monks Wood Experimental Station
Abbots Ripton
Huntingdon, Cambs PE17 2LS

Introduction

It is appropriate that both the Natural Environment Research Council (NERC) and the Biological Records Centre (BRC) should be celebrating a 25th Anniversary in the same year. Maintenance and development of long-term databases, such as are held by BRC, are central to many scientific programmes in NERC. It is also appropriate that the Conference organised to celebrate the two Anniversaries should be held at the Linnean Society. The Society has played a central role in leading discussions on the co-ordination of biological recording by all the agencies, organisations and societies involved in data collection, storage and interpretation.

In Britain, we are fortunate that there are many thousands of experienced and expert 'amateur' biologists who are interested in recording changes in wildlife. For 25 years, the Biological Records Centre has provided an interface between these 'amateurs' and the professional users of information – government and voluntary wildlife conservation bodies, environmental planners, and the scientific community. Support for biological recording has been at the core of collaboration between NERC and the Nature Conservancy Council (NCC) as the databases provide essential information for assessing and monitoring changes in British wildlife. The databases held by BRC have been used by academic and Institute scientists in NERC for a variety of purposes. Knowledge of biodiversity is essential for developing programmes to maintain genetic resources. Long-term datasets on trends in species distribution are essential for assessing the effects of land use change and pollution.

In the first section of these proceedings, we learn how the *Atlas* project of the Botanical Society of the British Isles was formed in 1954, and how its basic concepts have been applied over the last 36 years. The formation of the BRC in 1964 is described, along with the pioneering developments in data processing and map production. The BRC has now been integrated into the Environmental Information Centre at ITE Monks Wood and will benefit from the application of modern computing techniques for data processing and mapping. The first section also includes a review of progress towards the integration of environmental data management, both in the UK and in a broader European framework.

In the second section, the four case studies of the effects of environmental changes on wildlife draw heavily on data from national recording schemes organised in collaboration with BRC. These papers illustrate that biological records, especially when utilised with other environmental data, are essential for ecological and biogeographical studies of the effects of land use changes and pollution. Time-series data on species sensitive to climate variables will become critical in predicting the consequences of global warming. Information on the rate and extent of spread of introduced species will be essential in assessing potential effects on native species and communities of importance in wildlife conservation.

The first paper in the third section reviews the importance of biological recording in establishing the distribution of species and communities of conservation value. This information has been central to the development of nature conservation legislation and practice. The second paper reviews the overall development of environmental policies in the UK. We have only to look at current environmental issues – acid deposition, land use changes, and global warming – to see the intense interest of the media and general public in sustainable development which maintains wildlife resources. It is clear that biological recording will continue to be essential for assessing the effects of changes in environmental legislation.

The Conference took place during the uncertain period following announcement of the Government's plans to reorganise the NCC. Following extensive discussions initiated by the Linnean Society, NERC has launched a Co-ordinating Commission for Biological Recording. It is essential that these activities should build on the BRC experience in developing collaborative projects for biological recording. Many new opportunities are opening up for the community involved in biological recording. The introduction of computerised systems for interactive processing, integrating and mapping datasets will revolutionise the speed at which information can be interpreted and disseminated. In addition, the growing need for a European perspective, and the increasing importance of global environmental issues, will strengthen the pressure for international integration of biological recording.

Professor T M Roberts
Director
Institute of Terrestrial Ecology (South)

Biological recording – past, present and future

BSBI distribution maps scheme – the first 40 years

F H Perring
24 Glapthorn Road, Oundle, Peterborough, Cambs PE8 4JQ

I am delighted and honoured to have been allowed the privilege of contributing on this splendid occasion, part of which is to celebrate the 25th Anniversary of the formation of the Biological Records Centre (BRC) at ITE Monks Wood. I am sure it is a source of great satisfaction to all those who were involved in the creation of BRC that it has continued and that, at last, its future seems to be assured. A dictionary of abbreviations at the time of its formation gave under 'BRC': British Reinforced Concrete – which indicates how tough and durable we had to be to survive, and British Rabbit Council – which I took as a reference to the fact that the 'buck stops here'!

One person who would be particularly gratified at the outcome is Max Nicholson, who, as Director General of the then Nature Conservancy, came to a meeting of biological societies at the Cambridge University Botanic Garden in the autumn of 1962 to suggest that such a Centre be set up, as it eventually was in April 1964.

Even those without a mathematical degree will appreciate that it is now nearly 26 years since BRC was formed: I shall not question the reason for this delay, but only remark that those responsible were wiser than they knew because this allows us to celebrate, at the same time, the Anniversary of another event which was, arguably, more important in the history of biological recording than the formation of BRC.

It is 40 years ago this month, in March 1950, that the Botanical Society of the British Isles (BSBI) held a Conference in London on 'The study of the distribution of British plants', the proceedings of which were published the following year (Lousley 1951). [It would have been nice to report that the meeting took place in these very rooms, but it was in fact held at the Royal Horticultural Society in Vincent Square.]

At the end of two days of papers which reviewed the state of knowledge of plant distribution in Britain, the difficulties of acquiring accurate information, how data could be most usefully presented, and ways in which the situation could be improved, Professor Roy Clapham gave the final paper which ended with the following paragraph:

> 'it is high time we had a set of distribution maps of British species, and that we ought to set about the task; that the maps, when produced, should be *comprehensive* and *accurate;* that they should be *available*, . . . by being of small scale (1 in 10 million) so that they can be printed four on each page of a single Medium Quarto volume; that the unit area should be the 10 km grid square, . . . that overlays should show ground over 1000 and over 2000 ft and also basic sub-strata . . . and certain historical data should also be supplied if possible.'

Following a lively discussion, Dr Max Walters moved the final resolution that the BSBI should set up a project to map the British flora, which was carried with acclamation.

These two people were, of course, to play an enormous part in the success of the distribution maps scheme. Professor Clapham became secretary of the Committee under Lousley's chairmanship, and Dr Walters was honorary director for the first five years. I wish I could say I remember the meeting well, but I cannot – I was not there. I joined the BSBI in 1952 and was not aware of the scheme until, on a memorable visit to hunt for ground pine (*Ajuga chamaepitys*) at Odsey on the Cambridgeshire–Hertfordshire border in August 1953, Max popped the question, and I accepted the invitation to become the Administrative Officer – a post I took up in April 1954. [Here perhaps I can interpolate a fascinating series of coincidences. Although the 1950 Conference began in March, the first day was Friday, 31 March, and Professor Clapham's proposal was made on 1 April – All Fools Day; I took up my post on 1 April 1954, and moved to Monks Wood on 1st April 1964.]

Another vital role which Professor Clapham played in the success of the maps scheme, along with T G Tutin and E F Warburg, was the publication in 1952 of *Flora of the British Isles* which, for the first time since the 19th century, gave the average botanist the means, in one volume, to identify all the species and subspecies he was likely to find: it set a common standard to which all contributors to the scheme could refer and which they could all afford.

The other vital factor, without which the scheme could not have been efficiently operated or completed on time, was the advent of data processing equipment using punched cards. Without such equipment – and ultimately the mechanical production of the maps – it would not have been possible to handle the 1.5 million records that the scheme collected, or to transform these into the 1700 distribution maps which were published in the *Atlas of the British flora* (Perring & Walters 1962) only eight years later.

The punched cards were seen to be essential, initially as a means of changing lists of species by squares to lists of squares by species, and for arranging the cards in a sequence which could be listed on a tabulator to make hand plotting easier and more accurate. But, as each 40-column species card carried a unique 10 km grid square reference in four holes, we asked the manufacturers, Powers Samas, whether those holes could be used to bring up dots in different places on a preprinted map in the tabulator. The problem was, in fact, solved remarkably quickly by Roy Smith, a Powers Samas engineer, and mechanical mapping was demonstrated for the first time at the BSBI's 4th Biennial Conference at Church House, Westminster, on Friday, 9 April 1954. The first species mapped in this way was traveller's joy (*Clematis vitalba*) based on data collected by Edgar Milne-Redhead whilst driving around the country.

David Allen (1986) recorded in *The botanists* that Powers Samas were subsequently able to interest Kayser-Bonder, the hosiery manufacturers, in using the system to map the sale of nylon stockings – a good example of a commercial spin-off from pure science.

The tabulator was remarkably robust and continued to be used at BRC for many years producing the maps for the *Critical supplement* (Perring & Sell 1968) before being retired as a 'museum piece' at the entrance to BRC – it was still there when I left in 1978. In all modesty, I had felt that it was worthy of a place in a museum and tried to interest the Natural History Museum to no avail, and there was a proposed computer museum in this country which foundered through lack of support. You will be glad to know that the prophets who were not honoured in this country are remembered in the USA, and the tabulator, and the card sorter, are exhibited in the Digital Computer Museum at Marlboro, Massachusetts.

I do not think those who proposed the scheme in 1950 or those of us who were involved in setting it up were fully prepared for the response to it by the BSBI membership, or indeed by the general public. Within two years we had received offers of help from over 3000 volunteers: a single article by John Gilmour in the *Observer* brought 800 replies. Extra staff had to be employed, indicating, perhaps for the first time, the enormous potential of the use of amateur naturalists as a volunteer labour force for collecting data of this kind – particularly in relation to common and easily identified species. Of course, we would approach the problem differently today and use the network of county recorders and local records centres who are in touch with the active field botanists in their vice-counties, but it should be

remembered that, in 1954, there were very large areas where no such infrastructure existed: the whole of Wales, apart from Monmouth, was covered by A E Wade from the National Museum of Wales in Cardiff; E C Wallace from London, had the whole of Argyll, Ross and Sutherland; and Professor Webb the whole of Ireland. It was the maps scheme which changed the situation – people became involved and committed and, when the field work for the *Atlas* ceased, continued to work in their own area. By 1964, the number of recorders in Wales had increased from two to 11; Wallace's six Scottish vice-counties each had its own Scottish recorder; and the 40 vice-counties in Ireland were shared amongst 17 botanists.

The work force was nevertheless suited to the major task recognised by the 1950 Conference. Whilst knowledge of the distribution of rare plants in Britain was adequate and could mostly be gathered from herbaria and literature, data on common species, which could provide an accurate assessment of the limits of distribution, were extremely sparse. This was emphasised for me when, as an experiment, I sat at the feet of the great 'amateur' N D Simpson, author of *A bibliographical index of the British flora* (1960), in his magnificent library in Bournemouth where I was able to abstract all the known data on the flora of Roxburghshire, vice-county 80. For 22 squares I abstracted about 1600 records, but one square, around Kelso, had 355 (nearly a quarter), ten squares had less than 25 records, and three had none. Amongst the 1600 records, we had the following: maiden pink (*Dianthus deltoides*) 8, moschatel (*Adoxa moschatellina*) 7, creeping ladies' tresses (*Goodyera repens*) 5, yellow star-of-Bethlehem (*Gagea lutea*) 3, cow parsley (*Anthriscus sylvestris*) 2, heather (*Calluna vulgaris*) 2, cock's-foot (*Dactylis glomerata*) 1, daisy (*Bellis perennis*) 1, and shepherd's purse (*Capsella bursa-pastoris*) 0. Even the Kelso square had only 100 of the 400 'common' species which are recorded from every vice-county in the British Isles.

The major interest in the distribution maps at that time was a biogeographical one, particularly the wish to study the relationship between climate, soils and altitude, outlined by Professor Clapham in 1950, made possible by the overlays provided with the *Atlas*. There was the additional interest in using the information to shed light on the climate of the past, based on the rapidly growing body of information from sub-fossil deposits coming out of Professor Godwin's Sub-Department of Quaternary Research in Cambridge, and which was elegantly demonstrated in a paper by Conolly and Dahl (1970). Soon after the move to Monks Wood, a research assistant of Professor Godwin's made weekly visits to BRC preparing data for the second edition of *History of the British flora* (Godwin 1975).

What is almost unbelievable to us now, perhaps, is that nowhere in the 1950 Conference report is there a reference to conservation – despite which the

Nature Conservancy provided 50% of the funding for four of the first five years, and all the funding for the second five-year period of the project. Yet, by the time the *Atlas* and *Critical supplement* had been published, the threat to our flora, particularly from modern agriculture, was only too apparent, and Monks Wood Experimental Station, to which we had now moved, had been set up in 1961 to study that threat to our flora and fauna. BSBI and BRC have particular reason to be grateful to Max Nicholson, who was at that time Director-General of the Nature Conservancy. It was far-sighted of him to ensure financial support from the Nature Conservancy for the *Atlas* over a nine-year period until it was completed (even though conservation was not a primary objective), and to help create BRC.

It was only then, I believe, that the full potential of the *Atlas* as a conservation tool was appreciated, though the fact that we had data of immediate value was only as a by-product of the method. With many of the data from the remoter areas, especially in Ireland and the Highlands of Scotland, being based on a visit to a 10 km square for one day or less, it was appreciated that rare species would be overlooked and under-recorded, but, on the other hand, these were the species best represented in herbaria and the literature.

Thus, it was determined that, for the rare species, the 'A' species in the *Atlas*, arbitrarily defined as occurring in 20 or fewer vice-counties, a thorough search of herbarium and literature sources should be made and prominent botanists consulted. This list covered over 400 species – about 25% of the total – and, because we had the sorting facilities, 'old' records (before 1930) were shown on the maps in the *Atlas* with a circle rather than a dot.

So it was that in the late 1960s, instead of settling down to write that book on phytogeographical relationships of the British flora for which Dr Walters and I had signed a contract, we found ourselves much more concerned with the current changes in our flora and the threats which faced it. A new member of staff was appointed at Monks Wood to carry out a survey of the up-to-date distribution of the 'A' species, the list now pruned to include only those occurring, post-1930, in 15 or fewer 10 km squares. The next Conference of the BSBI, in September 1969, the proceedings of which were published the following year (Perring 1970), was on the theme 'The flora of a changing Britain'. I gave a paper on the previous 70 years, which drew attention to the serious decline in many of our rare species, and Max Walters presented a paper on the next 25 years.

Those data were available for the first Conservation of Wild Plants Act 1975, and the work of survey and report has continued ever since as a combined operation involving the Nature Conservancy Council (NCC), BRC, BSBI members, and the Royal Society for Nature Conservation (RSNC). The organisations combined with the World Wildlife Fund to publish both editions of the *Red Data Book 1: Vascular plants*

in 1977 and 1983 (Perring & Farrell 1977, 1983), which have influenced Schedules of species now protected under the Wildlife and Countryside Act 1981.

Although there have been surveys of some obvious groups, like arable weeds where changes might be anticipated, the list of species being kept under surveillance is still based on the original 'A' species for the *Atlas* selected in the mid-1950s.

It was because the BSBI believed that we should periodically be looking at what is happening to the whole flora that, in 1986, the Society applied to NCC for funds for a three-year project called the monitoring scheme, which was seen as the start of a three-phase operation to produce a new *Atlas of the British flora* in the late 1990s.

The scheme had two aims:
–   to make a comparison between the flora recorded in the period 1954–59 for the *Atlas* with the flora recorded today in an 11% sample of 10 km squares (1 in 9);
–   to list, within each of the same 10 km squares, three 2×2 km squares (tetrads) which would be the basis for monitoring future change in the flora.

The application to NCC was successful, as was the scheme under the dynamic drive of Dr Tim Rich. By one of those coincidences of timing which can only point to a divine co-ordinator, the completed typescript of the report was delivered to the chairman of the monitoring scheme committee yesterday (12 March 1990); it is likely to be received by the committee on 27 March. Despite 1 April 1990 falling on a Sunday, it would be only proper, it seems to me, if someone from the NCC in Peterborough could be present at Northminster House to receive the report they commissioned, exactly 40 years after Professor Clapham's original proposal.

It is one of life's little ironies that now the true value of the *Atlas* as a conservation tool is widely recognised, the Nature Conservancy Council has been unable to fund the BSBI's application to continue with the next phase of the project to publish a completely new edition in about seven years' time.* In the meantime, we have to continue with the old *Atlas* which is about to be reprinted yet again (Perring & Walters 1990).

REFERENCES

Allen, D.E. 1986. *The botanists*. Winchester: St Paul's Bibliographies.

Clapham, A.R., Tutin, T.G. & Warburg, E.F. 1952. *Flora of the British Isles*. Cambridge: Cambridge University Press.

Conolly, A.P. & Dahl, E. 1970. Maximum summer temperature in relation to the modern and quaternary distributions of certain arctic-montane species in the British Isles. In: *Studies in the vegetational history of the British Isles*, edited by D.Walker and R.G. West, 159–223. Cambridge: Cambridge University Press.

* *Editor's footnote*. Subsequently, NCC has placed two staff, on two-year contracts, at BRC, to assemble a database on Britain's scarce plants. The project began in October 1990.

Godwin, Sir H. 1975. *History of the British flora*. 2nd ed. Cambridge: Cambridge University Press.

Lousley, J.E. ed. 1951. *The study of the distribution of British plants*. Arbroath: Botanical Society of the British Isles.

Perring, F.H., ed. 1970. *The flora of a changing Britain*. Hampton, Classey.

Perring, F.H. & Farrell, L. 1977. *British Red Data Books: 1, Vascular plants*. 1st ed. Lincoln: Royal Society for Nature Conservation.

Perring, F.H. & Farrell, L. 1983. *British Red Data Books: 1, Vascular plants*. 2nd ed. Lincoln: Royal Society for Nature Conservation.

Perring, F.H. & Sell, P.D. 1968. *Critical supplement to the atlas of the British flora*. London: Nelson.

Perring, F.H. & Walters, S.M. eds. 1962. *Atlas of the British flora*. London: Nelson.

Perring, F.H. & Walters, S.M. eds. 1990. *Atlas of the British flora*. Revised 3rd ed. Oundle: Botanical Society of the British Isles.

Simpson, N.D. 1960. *A bibliographical index of the British flora*. Bournemouth. Privately published.

The Biological Records Centre: a pioneer in data gathering and retrieval

P T Harding and J Sheail

Institute of Terrestrial Ecology, Monks Wood Experimental Station, Abbots Ripton, Huntingdon, Cambs PE17 2LS

INTRODUCTION

The establishment of the Nature Conservancy in 1949 brought unprecedented opportunities for the study and conservation of plant and animal species, and their communities (Sheail 1987). How well equipped were botanists and zoologists to discharge these responsibilities? Could they provide the kind of national perspective that would be required to determine priorities? Captain Cyril Diver, the Director-General of the Conservancy, wrote to the Botanical Society of the British Isles (BSBI) in March 1950, emphasising how it would be impossible to fulfil all the duties laid on the Conservancy without the full support and co-operation of the national biological societies. They were of obvious value in carrying out biological surveys, and in acting as a bush-telegraph in reporting threats to 'smaller species sites'.[1]

This paper outlines how that collaboration was developed, first in the form of the BSBI plant mapping scheme and then, under the aegis of the Nature Conservancy (and latterly the Natural Environment Research Council), at the Biological Records Centre, located at Monks Wood Experimental Station.

THE ATLAS OF THE BRITISH FLORA

In 1938, Diver had discussed the need to map the distribution of species, as a means to understanding the limiting factors in the ranges of species, in the context of a grandiose national atlas (Taylor 1940). However, the stimulus to Diver's 1950 correspondence was the decision to devote the BSBI Conference of 1950 to the 'Aims and methods in the study of the distribution of British plants'. The absence of any kind of national overview of the distribution of every plant species was an obvious embarrassment. Whilst much might be gleaned from the published literature and museum collections about rare plants, detailed and systematic field surveys would have to be organised to achieve the same level of detail for the more common species. A model already existed in the form of Eric Hultén's *Atlas of the distribution of vascular plants in northwest Europe* (Hultén 1950). The adoption of a National Grid by the Ordnance Survey opened up the possi-

bilities of using, say, the 10 kilometre grid square, which appeared on every Ordnance Survey map, as the basic mapping unit for all species (Figure 1), whatever their degree of rarity.

The Conference concluded with a paper given by Professor A R Clapham, in which he put forward the idea of publishing an accurate, up-to-date and detailed atlas of British vascular plants, available for purchase in a convenient form and at a reasonable price. Whilst the biological vice-county system developed by Watson and Praeger had the advantage of being a familiar and established mapping unit, Clapham (1951) illustrated how its use often gave a misleading impression as to the continuity or otherwise of the distribution of many species.

A resolution encouraging the Council of the BSBI to 'discuss the possibility of preparing and producing a

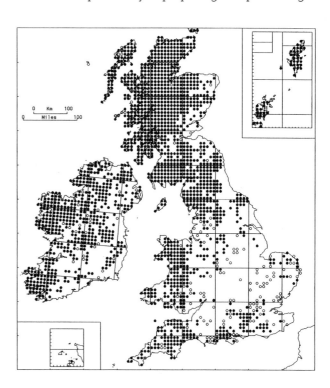

● Post-1950

○ Pre-1950

Figure 1. Distribution of *Drosera rotundifolia* showing the recorded occurrence of the species summarised by 10 km squares of the British and Irish National Grids

[1] Superscript numbers refer to manuscript sources – see p 19

series of maps of the British flora', was carried with acclamation, and a Maps Committee was formed. An approach to the Nuffield Foundation for a five-year grant of £10,000 was successful, and in April 1954 work began on mapping the distribution of British flowering plants and ferns, directed by Dr S M Walters and, from March 1959, Dr F H Perring, of the Botany School, Cambridge. Over 1.5 million records were contributed by about 1500 professional and amateur and volunteer workers. Records were received from all but seven of the 3651 10 km squares covering the whole of the British Isles. With the aid of data processing machinery, it proved possible to use punched cards, not only to collate the data for each species, but to make the final maps on a slightly modified commercial tabulator. It took less than one hour to place all the dots for a single species in their correct positions on the map (Nature Conservancy 1962; Perring & Walters 1962).

The original intention was to complete the collection of data by 1959, and for the maps to be ready for publication a year later. To the consternation of the Nature Conservancy, which met 70% of the eventual costs of £36,000 of the project, two extensions were required. In a sense, the project was a victim of its own success. More time was needed to edit the number of records which was much larger than expected. Both Clapham and Professor H Godwin of the Botany School, Cambridge, insisted that continued support should be given, not least because the project had reached a stage when the 'careful analytical and experimental research' could begin. Already the incidental publication of fragments of work in support of particular research investigations had attracted much attention, both in Britain and on the Continent. The Conservancy would gain much credit from association with the Survey.[2]

The *Atlas of the British flora* (Perring & Walters 1962) was eventually published for the BSBI by Thomas Nelson in 1962 at a cost of 5 guineas. Measuring 12 inches by 10 inches, it contained 406 pages of maps (four maps to the page) showing the distribution of about 1700 species. Intended as a factual document of sufficient accuracy to make it 'a valuable tool for biologists and all others whom it may happen to interest', the *Atlas* provided, for the first time, detailed information about the distribution of British plants on a uniform scale. It was also a pioneer in the use of data processing machinery for dealing with records at all stages, from their receipt until they appeared as symbols on a map.

THE FUTURE OF THE BSBI MAPPING SCHEME

Long before the survey was finished, recorders had begun to use the field record cards for a range of different purposes. Other organisms and areas might be mapped in a similar manner. The BSBI was, however, exceptional in being able to collect so many data. A total of 250 000 records a year had justified expenditure on data processing equipment. Few

other bodies could support individually the levels of investment and organisation that would be required. In their submissions, supporting the second and final extension of grant aid, Clapham and Godwin emphasised how every effort should be made to avoid disbanding a survey so competently directed, and which had several years' experience and the goodwill of a large band of voluntary helpers. A permanent mapping centre should be established as part of 'the scientific equipment of the country'.[3]

At a meeting convened in Cambridge in November 1962, with Max Nicholson (Director-General of the Nature Conservancy) in the chair and representatives of all the main biological societies present, there was a majority strongly in favour of setting up such a permanent centre. Its first priority would be to advise on the methods of studying the population and distribution of organisms so as to build up a record of their diversity, abundance and range. The second would be to process and edit the information received, so as to make it available to all amateur and professional biologists (Nature Conservancy 1964).

By taking responsibility for such a centre, and providing a service in data processing and map production, the Nature Conservancy would acquire, in exchange, an invaluable store of information on the distribution of British plants and animals. At its meeting in July 1963, the Nature Conservancy committee agreed to the transfer of Perring and his staff to the recently opened Monks Wood Experimental Station, near Huntingdon.[4] The removal took place in April 1964 and, three years later, the unit became a fully integrated part of the Conservancy, which, by that time, had been absorbed as a Charter Committee of the Natural Environment Research Council.

BIOLOGICAL RECORDS CENTRE, 1964–1973

Progress reports were submitted to the biannual meetings of an Advisory Sub-Committee, responsible to the Conservancy's Scientific Policy Committee. Clapham (who had been secretary of the BSBI's Map Committee) was chairman until 1968, when he was succeeded by Dr H C Gilson. At its first meeting, ' no positive agreement' was reached as to what the centre should be called. The word 'Information' was rejected as too general. Biological Data Centre was considered, but, at the second meeting, Perring's suggestion of the Biological Records Centre (BRC) was accepted.[5]

The meeting of the Advisory Sub-Committee in October 1966 concluded with a discussion as to how the work of the Centre might attract more publicity. A small silhouette of a frog was adopted as a motif on cards, maps and notices. The first Newsletter was issued in April 1967.[6] It was not long before the BRC was one of the best-known components of the Nature Conservancy. As well as visitors to the Centre, staff seemed to spend more and more time away, discussing and lecturing about matters concerning recording.

As anticipated, a high priority was given to processing the data provided by national biological societies.[7] The British Bryological Society was one of the first societies to use the Centre's expertise. Every October, records would be received, from which the next set of maps was prepared for the *Transactions of the British Bryological Society*. Maps of nearly 150 species were published over the period 1963–71. The British Conchological Society and BRC published the first *Atlas of freshwater and terrestrial Mollusca* in 1976.

The obvious incompleteness of the maps produced for the various societies had a stimulative effect. Some 500 additional records were received, following the publication of a set of interim maps for the British Deer Society. Some groups were so difficult to identify in the field that the production of a guide or key was essential. As organiser, Dr D V Alford produced a series of such guides for the 300 recorders who took part in a bumblebee distribution map scheme launched by the International Bee Research Association. By the early 1970s, the volume of work carried out for the British Trust for Ornithology (BTO) and Wildfowl Trust was so great as to justify each having a member of staff outposted to BRC. From 1965 onwards, the Centre was responsible for determining the totals of the number of birds of each species ringed and recovered under the BTO schemes. Once verified, the 80-column punched cards were duplicated, a set being kept at the Centre for analysis. The other set was sent to Paris, where all the European bird-ringing data were collated. In February 1968, a scheme was launched to produce an *Atlas of breeding birds* by the British Trust for Ornithology and the Irish Wildbird Conservancy, using the 10 km square as the basis for recording.[8]

Botanical recording

Following the publication of the *Atlas of the British flora*, two major tasks remained. For what were called 'technical and tactical reasons', the *Atlas* had omitted maps of a number of difficult genera, including *Alchemilla, Euphrasia, Hieracium, Sorbus* and *Rubus*. These genera were included, together with a number of interesting examples of maps of subspecies, varieties and hybrids, in a *Critical supplement*, eventually published in April 1968 (Perring & Sell 1968).[9] Sales compared well with those of the original *Atlas*, with 955 copies, representing 40% of the print run, sold in the first six months. The 500 maps were accompanied by taxonomic notes, and phytogeographical comments on each taxon or group of taxa, as well as an account of how each map was compiled.

The second urgent task was to review the status of the 400 rarest species in the British flora which had been identified by the letter 'A' in the *Atlas*. Two sets of lists were prepared. The first set comprised those species found in fewer than nine 10 km squares, and the second those occurring in nine to 15 10 km squares. The lists, comprising 300 species in all, were distributed in October 1967 and 1968 respectively to the recorders of the BSBI, the Conservancy's regional staff, and other interested persons, who were asked to provide up-to-date information on each population (Perring 1970). The detailed replies obtained from the majority of respondents, and any further records, formed the basis of a new set of maps that were used in the preparation of scientific evidence in support of a Wild Plants Protection Bill in 1968. However, the Bill was not passed by Parliament (Perring & Farrell 1977).

At last, there was a more rigorous basis upon which to assess the increase or, more usually, decline of species (Table 1). At a Conference organised by the BSBI on the 'Flora of a changing Britain', Perring (1970) described how a third of the localities of the 278 rarest species had been destroyed. About 7% of the native flora of about 1500 species might be threatened with extinction. Priority was given to the preparation of the British equivalent to the *Red Data Book* produced by the International Union for the Conservation of Nature, listing all the rare and declining species of flowering plants and ferns, and giving information about their former distribution, their present rate, and possible causes for decline. Such a book would alert the conservation movement and landowners to the precarious state of some species, particularly those 80 species now confined to only one or two populations. However, the *Red Data Book* was not published until 1977 (Perring & Farrell 1977).

Table 1. Number of squares and localities for the 278 rarest British plant species (Perring 1971)

	Squares	%	Localities	%
Historical period	3390	100	4595	100
1930–50	1673	49	1902	41
1960 onwards	1176	35	1425	31

Zoological recording

Right from the start, the question was posed as to how far BRC should go in initiating its own surveys. There were some groups, and most notably insects, where no suitable society existed to undertake a mapping scheme. Zoological recording was started at BRC as soon as the Centre was set up, and a common species recording scheme was launched in 1967 to cover 19 groups of animals (excluding birds and marine organisms). The scheme, which covered 530 species that the general naturalist could be expected to record with accuracy, never prospered and was soon abandoned.

It was not until John Heath joined BRC in 1967 that recognised zoological expertise was resident at BRC. As had been the case with botanical recording, it was essential that the zoological staff at BRC would be closely involved with national and regional societies. Heath was a well-respected amateur lepidopterist who had many contacts with amateurs and professionals in the British Isles and abroad. His approach to organising the collection of data for many groups of insects (and other invertebrates) was characteristically optimistic and ambitious.

The Lepidoptera distribution maps scheme was launched in 1967 with an advisory committee of five external professional and amateur lepidopterists, chaired by Sir George Varley. This committee met three times before transforming into a similar role for the insect distribution maps scheme, but it seems not to have met after December 1969. In a progress report for 1971, Heath (1971) listed the following groups covered by the insect distribution maps scheme: Lepidoptera, Odonata, Orthoptera, Hymenoptera – Bombidae and Formicidae, Siphonaptera, and Diptera – Tipulidae. By this time, BRC was also involved with national schemes covering vertebrates and various non-insect invertebrate groups (Table 2).

The Lepidoptera scheme got off to 'a very encouraging start', with publicity in newspapers, journals, and through societies and exhibitions. By the end of 1968, 900 recorders had returned 3700 field cards and 25 000 individual record cards from 1200 10 km squares. As early as 1970, BRC was able to publish part 1 of the *Provisional atlas of the insects of the British Isles* (Heath 1970) which covered the native butterflies, and in 1973 part 2, covering 100 species of larger moths, was published (Heath & Skelton 1973).

The international dimension

Right from the start, the horizons of the BRC were worldwide. Under the aegis of the United Kingdom's contribution to the International Biological Programme (Conservation of Terrestrial Communities), the BRC took part in a Check Sheet Survey to collect information on sites of biological importance in the world. The intention was to estimate how far major natural and semi-natural ecosystems were adequately protected, and what further steps were required to meet that end.

Plants

An exhibition of data processing equipment, organised by Perring, at the International Botanical Congress in Edinburgh in 1964, drew attention to the very ambitious mapping projects that could now be contemplated. A scheme to map the whole of Europe had become more feasible. The imminent publication of the first volume of *Flora Europaea* would provide a common basis for nomenclature and taxonomy (Tutin *et al.* 1964). The availability of a complete set of maps for Europe on a 1:1 000 000 scale, bearing the Universal Transverse Mercator Grid, meant there was a common mapping unit, especially as the same grid was also being used on large-scale maps within individual countries (Perring 1967).

Table 2. A chronological list of the national biological recording schemes operated by BRC in 1990

Schemes established before BRC	Date of origin
Vascular plants	1954
Myxomycetes	1957
Siphonaptera	before 1960
Coleoptera: Atomariinae and Ptiliidae	before 1960
Bryophytes	1960
Non-marine Mollusca	1961

Schemes established 1964 to 1973	
Amphibians and reptiles	1964
Aculeate Hymenoptera	1964
Butterflies (and Macro-Lepidoptera)	1967
Spiders	1968
Orthoptera, Dermaptera, Dictyoptera	1968
Odonata	1968
Non-marine Isopoda	1968
Pseudoscorpiones	1970
Diplopoda	1970
Chilopoda	1970
Coleoptera: Elmidae	1970
: Staphylinidae	1970
Diptera: Tipulinae	1970
: Dixidae	1970
: Sciomyzidae	about 1970
Mammals	1971
Marine algae	1971
Coleoptera: Coccinellidae	1973

Schemes established 1974 to 1990	
Opiliones	1974
Coleoptera: Carabidae	1974
Tricladida, freshwater flatworms	1976
Diptera: larger Brachycera	1976
Cladocera	1977
Neuroptera, Mecoptera, Megaloptera	1977
Diptera: Syrphidae	1977
: Conopidae	1977
: Sepsidae	1977
Characeae	1979
Freshwater Oligochaeta	1979
Hemiptera: Auchenorhyncha	1979
Coleoptera: aquatic beetles	1979
Trichoptera	1980
Coleoptera: Chrysomelidae and Bruchidae	1980
: Scolytidae	1980
Diptera: Culicidae	1981
Lepidoptera: Incurvarioidea	1982
Coleoptera: Elateroidea	1982
: Cerambycidae	1982
Hemiptera: aquatic Heteroptera	1983
: terrestrial Heteroptera	1984
Gasteromycetes	1985
Coleoptera: Scarabaeoidea	1986
: Cleroidea, Lymexyloidea and Heteromera	1986
: Cantharoidea and Buprestoidea	1986
: Curculionoidea (part)	1986
Tricladida: terrestrial flatworms	1987
Hymenoptera: Symphyta	1987
Plant galls	1989

A working party was set up, with a view to compiling trial maps for ten species, using the 50 km squares of the Universal Transverse Mercator Grid as the mapping unit. Progress was sufficiently encouraging for the decision to be taken, during the 4th Flora Europaea Symposium in Denmark, in August 1965, to proceed with mapping all the species included in the *Flora Europaea*. The secretariat was to be located in Helsinki, and Perring was elected to the co-ordinating committee. The first volumes covering the Pteridophyta and the Pinaceae to Ephedraceae were published in 1972 and 1973 respectively[10] (Jalas & Suominen 1972, 1973).

Invertebrates

At a meeting in Paris in October 1967, Professor Jean Leclercq read a paper to the Société de Biogeographie, in which he set out a proposal to map the insects of Europe. At the same meeting, Heath described the data collection and processing methods used by the BRC. A meeting at Monks Wood in the following March, attended by four members from Leclercq's department at Gembloux in Belgium, agreed on the design of field cards and an 80-column individual records card for a mapping scheme that would use the base maps already available from the Flora Europaea mapping scheme. Preliminary notices of the European Invertebrate Survey were published and sent out by direct mailing, the BRC and Gembloux meeting the costs jointly.[11] The first invertebrate symposium was held in Saarbrucken in July 1972, and a second in August 1973 at Monks Wood, at which 62 delegates from 23 countries adopted a constitution and appointed a permanent committee, with Heath as secretary-general (Heath 1975).

Technological changes

Whilst much of the Centre's attention had to be given to co-ordinating the collection of data, it was equally important that the information, once received, was handled, stored and retrieved efficiently. Not only were distribution records and analysis tending to become more complex, but the number of enquiries increased.

The specially modified ICL 40-column tabulator, with associated punch and sorter, using a base map specially drawn to fit the tabulator, had proved an inexpensive and convenient system for plotting the maps of the botanical *Atlas*. At its second meeting, the Advisory Sub-Committee recommended the purchase of 80-column equipment, including an interpreting punch and a pattern select sorter. However, by the time this further equipment was installed, it was clear that the days of such a system were numbered. As a letter from ICL made clear, it would soon be impossible to obtain maintenance and spare parts as the machinery become obsolete.[12]

The benefits to be gained from adopting a computer-based system were obvious. Once the basic records were held in computer store, updating could take place continuously and, with suitable subroutines, it would be possible to map associations between two or more species, or list those occurring in a single square or group of squares. To oversee the conversion of the system based on 40-column cards to a fully computerised system, Miss Diana Scott, from the Royal Aircraft Establishment, was appointed as Data Processing Officer. In 1970, a Teletype computer terminal was installed, linked by telephone line to the Atlas-2 computer at the Computer Aided Design (CAD) Centre at Cambridge. The first stage of the conversion to computers involved the transfer of 1.5 million existing plant records from the 40-column cards to magnetic tape. The cards had to be transported to London in batches of 300 000, where the data were transferred on to magnetic tapes at the Law Society's Computer Centre. These tapes were then copied by London University's Computing Services on to the special tapes used by the Atlas computer. They were then rewritten in a condensed form, and indexed for easy retrieval of the data. The last batch of cards was dispatched for processing in June 1971.

Where might technical innovation end? As Perring (1970) remarked, it was important not to allow the sophistication of the machinery to blind users to the primary objectives of collecting data. Experience over 15 years indicated that nearly all the questions asked of the databank were either species- or locality-orientated. Irrespective of what else the technology might be used for, it was essential that these questions could be answered easily and cheaply.

BIOLOGICAL RECORDS CENTRE 1973–1989

In 1973 the Nature Conservancy was split to form two independent organisations, the Nature Conservancy Council (NCC) and the Institute of Terrestrial Ecology (ITE) (Natural Environment Research Council 1973). The separation of what were ostensibly the conservation and the research branches of the former Nature Conservancy created an inevitable problem over the most appropriate location for the Biological Records Centre. After a period of some uncertainty, BRC remained at Monks Wood, as part of the new ITE, within the Natural Environment Research Council.

In the general restructuring of funding systems in the Civil Service (the Rothschild Principle), the new Institute was commissioned to conduct research on contract to NCC. One of the areas of commissioned work was the BRC programme. Every year since 1974, NCC has supported the general operation of BRC with contract funds, although the percentage of the full costs of BRC covered by NCC's funding has steadily decreased over the years. However, in addition to direct funding to ITE to support the work of BRC, NCC has itself employed staff to work at BRC on specific projects, such as the BSBI monitoring scheme and the bryophyte atlas project. The contract between NCC and ITE supporting BRC recognises that the Centre is a joint responsibility to which both organisations allocate resources.

Since 'the split' in 1973, the work of BRC has changed and developed in response to altered circumstances and priorities, rapid developments in technology, and changes in staff. BRC has also undergone a succession of administrative changes since 1973 as ITE and NERC have developed new management structures. In 1989 BRC became the largest component in ITE's Environmental Information Centre (Wyatt 1992).

In the period immediately after 1973, BRC effectively lost its previously developing role as a data processing unit for the Nature Conservancy and, by the end of the 1970s, BRC was seen largely as a species distribution mapping unit. Partly in response to requests from NCC for information on species occurring at sites, such as National Nature Reserves and Sites of Special Scientific Interest, in 1981 a decision was made to incorporate more detailed site information into the computerised database (Table 3, Figure 2). Most recording schemes had already been collecting site-relatable information at least for the less common species. The early 1980s saw not only an intensive phase of computerisation of new data, in a site-relatable form, but also a renewal of contact with recording schemes to encourage them to record in more detail and for purposes additional to the publication of national distribution maps. In particular, the foundations were laid for the use of the BRC database as a tool for exploring biogeographical associations and, ultimately, to tackle the challenge of predicting changes in species populations and ranges in response to environmental changes.

These developments took place at a time when several new staff were joining BRC: Dorothy Greene and Paul Harding in 1979, Chris Preston in 1980 and Brian Eversham in 1983. Franklyn Perring had left BRC at the end of 1978, to be succeeded by John Heath. He retired in January 1982, and Paul Harding became head of the Centre.

Botanical recording

The large resource of data on vascular plants (approximately 1.5 million records) which formed the nucleus of BRC in 1964 has been augmented and updated in largely opportunistic ways, particularly during the last ten years. In addition, two new projects (BSBI monitoring scheme and the database

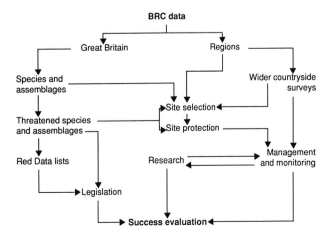

Figure 2. The application of site-relatable data, held by BRC, to nature conservation

Table 3. Biological Records Centre – data fields

Site relatable data (almost all data computerised since 1981)

Taxonomic	1.1	Order/genus/species/infra-specific taxon
Geographical	2.1	Country
	2.2	Vice-county
	2.3	Grid reference (10 km, 1 km, or 100 m square)
	2.4	Locality
	2.5	Site status (eg NNR/SSSI/NT/ Trust NR)
Temporal	3.1	Date (day/month/year or date period)
Personal	4.1	Recorder/collector
	4.2	Determiner
	4.3	Record compiler
Other	5.1	Altitude
(optional)	5.2	Habitat (land use/vegetation type, microsite)
	5.3	Record source (field/museum/ literature)
		5.3.1 Location of voucher material
		5.3.2 Reference to bibliography
	5.4	Species status (native, naturalised, etc)

and atlas of aquatic plants) have added to the database in a more directed fashion. Work on other botanical groups has involved myxomycetes, marine algae, lichens, charophytes and bryophytes, with new databases being established at BRC for all but the lichens.

Vascular plants – updating species
Many of the records contributed to the BSBI *Atlas* provided detailed information on localities, recorders and dates, and there has been a rolling programme to incorporate these more detailed records into the BRC database. Most of the species concerned are uncommon, localised or taxonomically difficult. This work has been expanded to incorporate more recent records from specialists and vice-county recorders, from recent publications, especially local floras, and from herbaria.

Requests for updated maps or data for individual species, for publications such as BSBI Handbooks and the British Ecological Society's *Biological Flora* series, or for research and ecological assessment, have justified the allocation of resources to updating the database for selected species. By 1989, maps of about 420 species and hybrids had been published in this way (Preston 1990).

An ambitious project to prepare a multi-volume *Flora of Great Britain and Ireland* was proposed by a consortium of experts in the early 1980s, and updated maps were prepared at BRC for most of the species expected to be included in the first volume of the *Flora*, before the whole project was abandoned by the originators.

Vascular plants – Red Data Book species
The maps published in the *Atlas of the British flora* drew attention to both rare and decreasing plant

species. From 1968 onwards, BRC made special efforts to encourage BSBI vice-county recorders to collect and submit localised records of such species. The data obtained were summarised in the first *British Red Data Book* (Perring & Farrell 1977, 1983), and a separate dataset on rare vascular plants has subsequently been maintained by NCC and BRC.

Vascular plants – BSBI monitoring scheme
Data summarised in the *Atlas of the British flora* are at least 30 years out of date. Dots in the *Atlas* show post-1930 or, at best, post-1950 records. New editions of the *Atlas* (in 1976, 1982 and 1990) have done little to remedy the lack of recent information, except for about 320 rare taxa.

As part of a phased project to compile a completely new and updated atlas, the BSBI initiated a study of more recent changes. This study surveyed a sample of one in nine of the 10 km squares of the National Grids in Britain and Ireland (Ellis 1986). The project was intended to identify species which are changing their distribution and to establish a baseline for monitoring changes regularly, independent of the proposed larger *Atlas* project. The project is described by Rich (1992). The scheme resulted in nearly 1 million records which have been added to the BRC database.

Vascular plants – database and atlas of aquatic plants
Work on the preparation of the BSBI Handbook on *Potamogeton* and allied genera started Chris Preston's interest in this neglected group of plants, and in aquatic macrophytes generally. In 1989 funding was obtained from the Water Research Centre to help support a new project to compile a database of aquatic plants, eventually leading to an atlas. The project is planned in two phases, the first 18 months being devoted to work on floating and submerged species, with the subsequent two years on emergent species.

Myxomycetes
Summarised (10 km square) data for selected species of myxomycetes (slime-moulds) were computerised for Bruce Ing and the British Mycological Society in 1980–81, and a *Provisional atlas* of 100 maps was published in 1982 (Ing 1982).

Marine algae
The British Phycological Society launched the marine algae recording scheme, in conjunction with BRC, in 1971. The resultant data were computerised in 1983–84 and, thanks to additional funding from NCC, a detailed site-relatable database was compiled at BRC incorporating records from NCC surveys. Distribution maps of a selection of 155 species were published in a *Provisional atlas* (Norton 1985).

Lichens
In 1963, the British Lichen Society launched its mapping scheme and subsequently became associated with BRC. In 1982, NERC published volume 1 of an *Atlas of the lichens of the British Isles* (Seaward & Hitch 1982) using hand-plotted maps provided by

the Society. No data for lichens have been deposited at BRC and the British Lichen Society now operates its mapping scheme independently of BRC (see Hawksworth & Seaward 1990).

Charophytes
Stoneworts have been neglected by British botanists in recent years, but Jenny Moore of the Natural History Museum has made a special study of the group. She collaborated with the BRC data manager, Dorothy Greene, to prepare a computerised catalogue of the Museum's collection and a *Provisional atlas* for all available records (Moore & Greene 1983). Updated maps were prepared by Dorothy Greene for a BSBI Handbook to the group (Moore 1986).

Bryophytes
The launching of the bryophyte recording scheme in 1960, by the British Bryological Society (BBS), predates BRC, but the scheme has been closely associated with the Centre since 1964. The methods used by BBS were modelled on those used by the BSBI for the *Atlas of the British flora*, but, with many fewer recorders, the collection of data inevitably took longer. Maps of 224 species were published in the BBS journal between 1963 and 1978 and a *Provisional atlas* (Smith 1978) covering 104 species was published by NERC.

The original scheme had been based on the compilation of 'master cards', holding summary data for each 10 km square, but by the mid-1980s technological advances and the requirement for site-relatable data encouraged BRC to input fully detailed records to the database. By the end of 1989, the 770 000 records collected by the BBS had been computerised and the first of three volumes of the *Atlas of bryophytes* (covering liverworts) is complete (Hill, Preston & Smith 1991). The preparation of a *Red Data Book* for bryophytes has been started by N F Stewart (of Plantlife, formerly the Conservation Association of Botanical Societies), using the database at BRC.

Zoological recording
Zoological recording started at BRC soon after it was set up in 1964, with schemes for reptiles and amphibia, and for mammals, being initiated, but invertebrate recording started 'from cold' in 1967 and within a few years many national schemes were in operation (Table 2). By 1989 there were 45 schemes covering nearly 10 000 species of terrestrial and freshwater invertebrates and with 1.25 million records of 3000 species compiled in the database. Work on vertebrates has always been seen as important and able to attract a strong (but sometimes inexpert) response from the general public. An *Atlas of mammals*, based on over 100 000 records, will be published in 1992. However, it is the macro-Lepidoptera (especially butterflies) which have been the cornerstone of BRC's work on invertebrates.

A summary of the progress of each of the zoological schemes or datasets is not practicable in this paper, but, by way of examples, we have selected three groups for their different approaches and results. A list of the zoological data holdings forms Table 4.

Butterflies

It has already been noted that maps of butterflies were published in 1970 and these maps were updated for publication in Howarth (1973). By 1982, when Heath retired, a large number of records had been contributed to the scheme and were summarised as 10 km square records and held on computer for mapping. As part of the move to compile site-relatable data on computer, it was decided to use the original records contributed by recorders to compile a new database (Harding & Greene 1984) which was summarised in the *Atlas of the butterflies of Britain and Ireland* (Heath, Pollard & Thomas 1984). That decision has resulted in what is certainly the most intensively used zoological dataset, with a steady flow of requests for data for research or survey of an individual species and for county-based surveys.

Table 4. Computerised zoological data holdings at BRC (1990)

Marine invertebrates

Dinoflagellates	27 200

Terrestrial and freshwater invertebrates

Tricladida (freshwater flatworms)	1 400
Hirudinea (leeches)	4 400
Opiliones (harvestmen)	11 200
Cladocera (water-fleas)	4 000
Amphipoda	500
Decapoda (freshwater crayfish)	2 300
Non-marine Isopoda	
Asellota (waterslaters)	500
Oniscidea (woodlice)	27 000
Diplopoda (millipedes)	17 000
Chilopoda (centipedes)	16 000
Orthoptera, Dermaptera and Dictyoptera	22 000
(grasshoppers and crickets, earwigs, etc)	
Odonata (dragonflies)	98 500
Rhopalocera (butterflies)	238 900
Macro-Heterocera (macro-moths)	394 500
Carabidae (ground beetles)	46 000
Staphylinidae (rove beetles)	8 600
Coccinellidae (ladybirds)	9 300
Aculeate Hymenoptera	
Apidae (social bees)	8 500
various (solitary bees and wasps)	900
Brachycera (horseflies, etc)	19 600
Syrphidae (hoverflies)	40 900
Dixidae (meniscus midges)	1 400
Dolichopodidae and Empidoidea	500
Sepsidae	6 000
Muscidae (cattle-visiting flies)	2 000
Non-marine Mollusca (slugs and snails)	144 000

Vertebrates

Freshwater fish	1 800
Amphibians and reptiles	2 300
Birds	284 000
Mammals	110 000
Total	1 551 200

The butterfly monitoring scheme launched in 1976 was previously independent of BRC, but in 1989 it became part of the BRC project and its database has now been incorporated with that of BRC. The work of the scheme is described by Pollard, Hall and Bibby (1986), and by Pollard (1992) and Yates (1992) in this volume.

Dragonflies

The Odonata were one of the groups covered by the original insects distribution mapping scheme set up in 1968. The scheme was slow to show results, but, by 1977, post-1960 coverage of about 20% of 10 km squares had been achieved and maps were published in a new field guide (Hammond 1977) and as a *Provisional atlas* (Heath 1978). Under a new volunteer organiser, David Chelmick, and with the new field guide, the scheme prospered and many new recorders were recruited. In 1981 the organisation of the scheme passed to Bob Merritt who, with dynamic enthusiasm and commitment, encouraged the scheme to collect detailed site-relatable records. The results of this scheme up to 1988 are summarised in the forthcoming *Atlas* (Merritt, Moore & Eversham 1992). Subsequently the scheme has concentrated on collecting even more detailed information in the key sites project, which aims to establish proof of breeding and to estimate numbers of individual species at sites.

Woodlice

Three recording schemes were launched in 1970 covering woodlice (terrestrial Isopoda), centipedes (Chilopoda) and millipedes (Diplopoda). These schemes set out to collect information on the habitat preferences of species, as well as detailed distribution data, using a hierarchical habitat classification designed specially for these soil and litter organisms.

With a core of 20 or fewer regular contributors, the woodlice scheme collected 27 000 records of 34 native and naturalised species over a period of 13 years. Identifications were carefully controlled by the organisers and, as a result of checking and returning identified specimens to recorders, regular newsletters and annual field meetings, it was possible to build up a team with experience in field craft and identification. The results of the survey were published with maps and habitat analyses for each species (Harding & Sutton 1985) and subsequent analyses have been undertaken (Sutton & Harding 1989; Harding *et al.* 1991).

The international dimension

As a pioneer in the collation of national biological survey data and in the use of computers in handling those data, BRC has had a role in the formation of comparable data centres in other European countries and elsewhere. That role has usually been passive, by providing a working example, with specialists from every continent visiting the Centre to examine its methods and facilities.

One of the stated objectives of the European Invertebrate Survey (EIS) had been to promote and encourage the establishment and activities of national centres for distribution studies of invertebrates and, in the period since EIS was formally constituted in 1973, data centres have been set up in several European countries. In reality, most of these centres owe their existence more to the efforts of individuals and national organisations than to any initiatives on the part of EIS.

Without a permanent secretariat or funding, EIS was unlikely to prosper in the increasingly difficult financial climate of the 1970s and 1980s. The organisation has survived, albeit in an altered form, and is best regarded as an *ad hoc* assemblage of specialists from individual countries with an involvement in species distribution studies and invertebrate conservation. Under the guidance of its current president, Dr M C D Speight, EIS has focused its attention on providing specialist advice on invertebrates to the standing committee to the Berne Convention (Council of Europe 1990).

For some years, the Council of Europe has provided a focus for thinking on databanks in wildlife conservation and on the need for standards, particularly in species nomenclature for legislative purposes in the context of cross-national information exchange. Three 'colloquys' around the theme of computer applications in nature conservation were held at Strasbourg, in 1983, 1985 and 1986, with Harding representing BRC at the last. A catalogue of databanks in the field of nature conservation was published (Council of Europe 1985); information was gathered in 1987 to update that catalogue, but was never collated or published. A select committee of the Council of Europe, chaired by Harding, advised on the need for a nomenclatural database of European vertebrates, as a first step towards a series of such databases (Council of Europe paper PE-R-BD (87) 3 rev.) (Harding 1990c). No further action has been taken on any of these aspects because the relevant section of the Council has been unable to allocate resources for either meetings or consultancies.

Ad hoc collaboration by specialists is still possible, but fails to realise the potential of international collaboration, because of a lack of resources. Projects such as *Atlas Florae Europaeae* continue, but largely as a result of the dogged persistence of a few activists. Only now is the *need* for a comprehensive overview of the European distribution of species being recognised by other scientists, particularly those concerned with research on the potential effects of climatic change.

The European Commission (EC) has yet to have an impact on biological recording although the CORINE project (Co-ordinated Environmental Information in the European Community) has shown that international data exchange is practicable in relation to biological information. As the EC takes a more active role in wildlife legislation (eg through the Habitats Directive and the proposed Environment Agency), it can be expected that the existence of national biological data centres, such as BRC, will provide a resource to be developed and utilised through international collaboration.

Technological changes

The period since 1973 has seen an almost universal explosion in the use and development of electronic data handling procedures, and, in particular, in the increasing improvement and sophistication of computers and software with resultant cost-effectiveness. The creation of ITE in 1973 saw a rapid expansion of scientific computing facilities throughout the Institute. A Digital PDP 11–10 computer was installed at Monks Wood in March 1974, and this offered new opportunities to update the data handling and map production capabilities of BRC.

However, it was not until the late 1970s that truly effective use was made of computerised data management technologies for both data banking and the production of output, especially distribution maps. The delay was due in part to the sheer size of the BRC dataset as a whole – it was too large to be handled efficiently on the PDP 11–10, although subsequent upgrades of the Monks Wood computer enabled work on individual datasets to be undertaken. Until 1978 all map production still relied on the modified IBM 870 Document Writing System, adapted to read 80-column punched cards, which was introduced in 1969. In 1977 the NERC-funded Experimental Cartography Unit collaborated with BRC to produce maps using their Laser-scan HRD2 high-resolution plotter, and the resultant maps (of 104 bryophytes) were published in a *Provisional atlas* (Smith 1978).

The close-down of the Atlas computer at the CAD Centre in Cambridge in 1977 led to data previously held there being transferred to a dual IBM 360/195 configuration at the Rutherford Laboratory of the then Science Research Council. For the first time, BRC data were held within a formal database management system – the G-EXEC package, developed by the Institute of Geological Sciences. By 1978, nearly 2 million records (including 1.35 million records of vascular plants from the provisional *Atlas* project) were managed under G-EXEC.

The period since 1978 has seen a steady development in the computing facilities available to BRC. The installation of a workstation in August 1979, connected to the renamed Rutherford Appleton Laboratory, enabled the BRC database to be accessed interactively and heralded a new phase in providing output to users in a variety of forms. In 1980 a new outline map of the British Isles was produced by NERC Computing Services staff and Barry Wyatt (then at ITE Bangor), which enabled the distribution of species to be plotted via the computer. The outline was first used to produce maps of sedges for a new handbook (Jermy, Chater & David 1982). The early/mid-1980s saw a period of computing stability during which a number of new datasets were added to the database and the complexity of data (in a site-relatable form) was increased (Table 3).

A further migration of the BRC database began in November 1986 as part of major rationalisation and upgrading of computers by NERC Computing Services, which included the adoption of the ORACLE relational database as the standard within NERC for handling its corporate databases.

The migration of datasets from G-EXEC to ORACLE was completed in 1989, thereby placing the BRC database in a modern and flexible data management environment. Later that year a local area network was installed at Monks Wood which linked microcomputers and terminals to a newly installed Micro-VAX cluster. Map production has been in-house at Monks Wood since 1989 using a laser printer (Figure 1).

The caution expressed by Perring (1970), that it was important not to be blinded by the sophistication of machinery, may have been justified at the time when computing was in its infancy. However, the equipping of BRC in recent years with appropriate computing power, output facilities, data management expertise and analytical methodologies has been crucial to the continued success of the Centre. It may seem surprising that it is only in its 25th year that BRC and ITE colleagues have been seriously able to address some of the biogeographical questions relating to species and to begin the 'careful analytical and experimental research' alluded to by Clapham and Godwin nearly 30 years earlier.[2]

Possible a new word of caution is needed: users of BRC's data should not be so blinded by the present opportunities to analyse and interpret existing datasets that the collection and processing of new data and the updating of existing data are neglected!

Local records centres

Formalised biological recording, organised at a local level (eg a county), took a new direction in the 1970s with several initiatives towards the formation of local records centres, mainly based at county or city museums (Stansfield 1973; Somerville 1977; Flood & Perring 1978). By 1980, there were at least 60 centres covering most English counties and parts of Wales and Scotland (Harding & Greenwood 1981). The coverage has not changed substantially in the last ten years, although a few centres have ceased to operate and others have been created or have become more securely established. Seventy centres were listed as being in operation in February 1987 (Berry 1988).

Although there is often close collaboration with BRC, there are no formal links with these local centres. As a result, no network or protocols for exchange of data exist, there is some duplication of effort, operating standards vary, and scarce resources are not being used to best effect. These problems were highlighted at a meeting convened by the Biology Curators' Group at Leicester in 1984 (Anon 1985)

and subsequently by the National Federation for Biological Recording (NFBR) (Copp & Harding 1985) and by a Working Party convened by the Linnean Society of London (Berry 1988). Progress towards national policies in biological recording, up to May 1989, was reviewed by Harding (1990d), and subsequently a Co-ordinating Commission for Biological Recording, chaired by Sir John Burnett, has been formed to provide a focus for work towards a national network in biological recording throughout the UK.

A recent initiative by NCC, vigorously supported by NFBR, and partly financed by the World Wide Fund for Nature through the Royal Society for Nature Conservation (RSNC), provides the most tangible hope for simplifying the exchange of biological records, between centres at all levels. The RECORDER data management package developed by Dr S G Ball of NCC was released early in 1991 and will enable users to manipulate their data in a carefully designed and extensively tested computer system. RECORDER will not solve all the problems of the interface between local centres and national bodies such as BRC, and, of course, NCC and RSNC, but it will enable data to be stored and exchanged in standard forms.

The proliferation of home computers in Britain during the second half of the 1980s has led to many individual recorders, some scheme organisers, and some local records centres computerising data 'at source'. Early trials in acquiring data in this form, rather than in the traditional form of record cards, have been successful, and we look forward to increasing amounts of new data coming to BRC already computerised and checked for accuracy. BRC will be able to incorporate new data in this form, with greater ease and speed, leading to more efficient use of the existing, finite resources of manpower and equipment available at the Centre. However, need for discipline in adhering to agreed standards is even more important when magnetic media replace paper.

THE USE OF BRC DATA

The steady accumulation of computerised datasets over the last 25 years has enabled BRC to expand the range of uses to which data are put. Given the location of BRC within a research council, the application of those data to environmental research is seen as an essential function. Although many of the contributors of data regard national species distribution maps as the main product of BRC, it is the database which is the principal end-product of their efforts. BRC data are now being put to a wide range of uses. Many of these uses had been recognised from an early stage, but there had previously been insufficient resources, particularly of detailed data, to realise the full potential.

Five factors are important in reviewing the range of uses to which BRC data are being and could be put.

1. Data are now held in an accessible relational

database management system – ORACLE – which allows flexible access by a variety of criteria and search conditions.

2. There is now a wide range of detailed and good-quality data.

3. As part of ITE's Environmental Information Centre, BRC has access to other environmental datasets, to sophisticated spatial data handling technologies and to statistical expertise.

4. The requirement for authoritative and comprehensive environmental data has increased greatly in recent years; in particular, there is a pressing need to document the consequences of past changes and to predict the course of change in the future.

5. In common with most government-funded research institutions, BRC must develop the commercial application of its data.

The uses to which BRC are now being put can be summarised under four headings: information, monitoring, ecological analysis, and conservation and evaluation.

Information

Information on the occurrence of species is not only the basic feedback required by the voluntary contributors of data, but also the most tangible product of BRC, especially in the form of distribution maps. Maps have been produced for over 7000 species, many of which have been published in *Atlases* and taxonomic guides (Harding 1989) or in papers and books (see, for example, Preston 1990).

With an historical perspective in the data, derived from published records and preserved specimens in collections, it is possible to examine whether the range of an individual species, or of an assemblage of species, has changed. Several of the following papers pursue this aspect, 'Changes in British wildlife' being the theme of this Conference, and we return to the topic ourselves later. Assessments of change, such as those made for a few groups of invertebrates, reinforce the need for the continued collection of data. The distributions of species are not static, and information about change is required to assess the effects of man's management of the natural environment and the effectiveness of any measures being taken to control rates of change and to conserve species and sites.

Information is required not only at the national (UK) level, but there is increasing need to be aware of our island flora and fauna in a European and even global context. European species mapping projects exist for vascular plants, birds, mammals, reptiles, amphibians and some invertebrates. At present these projects function on a voluntary basis and depend on the enthusiasm and commitment of the co-ordinators and collaborators. However, it is not unreasonable to speculate whether Europe needs more than these projects can provide, and whether international collaboration on species recording should not follow the lead on site recording and documentation of the

CORINE biotopes project described by Moss and Wyatt (1990).

Monitoring

Although BRC schemes are surveys and have no pretensions to monitor species or sites, the data collated by schemes provide baseline information on the occurrence of species. These data are then applicable to nationwide or regional projects to examine the effects of man on the natural environment, and the spatial and temporal components of BRC data can be used in analyses with similarly referenced datasets. In the case of the BSBI monitoring scheme (Rich 1992), the original data for the *Atlas*, collated in the 1950s, have been used in comparison with the recent data to examine changes, and the monitoring scheme itself provides a sample baseline for resurvey in future decades. The role of national species surveys in the context of monitoring for conservation is discussed by Harding (1990b).

As has been shown by the BSBI monitoring scheme, volunteers have proved themselves to be malleable to specific requests for information and to project-orientated surveys. The potential resource of experienced volunteers involved with national schemes, who could be directed to projects with a monitoring component, has yet to be exploited. Sampling frameworks are required which are representative of the national situation; the ITE land classification system (Bunce, Barr & Whittaker 1981) is one approach which offers opportunities in this context.

Ecological analysis

From the earliest days of the BSBI *Atlas* project, it was intended that the data from national distribution surveys would be used in analytical and experimental research. Little use was made of the BSBI *Atlas* data in any form of analysis although the data were used to provide a basis for the *Red Data Book* (Perring & Farrell 1977) and in the development of guidelines for the selection of Sites of Special Scientific Interest (NCC 1989).

Subsequently, data from some schemes have been used in a variety of analyses, in particular those schemes which were set up with objectives beyond obtaining an overview of the distribution of species, such as the three soil and litter fauna groups described earlier. The application of multivariate techniques to the analysis of BRC data has been developed particularly for water beetles and the ground beetles, as described by one of the posters (Luff *et al.* 1992).

Within ITE, analysis of BRC data is being directed in particular towards linking species data with other environmental datasets, such as climate, soils, land use, and habitat potential and availability. Such analyses are still at an early stage and are particularly concerned with modelling the possible consequences of climate changes. Other analyses are looking at patterns of species richness and the ranges of individual species and habitat-associated assemblages of species. One particular area of interest is the distribution of butterflies, where there are opportunities of linking data from the national recording scheme with those from the butterfly monitoring scheme.

Conservation and evaluation

The requirements for data from biological recording in wildlife conservation and in site evaluation have many points in common. In particular, localised information from site surveys needs to be put in a national context – for example, to assess the rarity or the degree to which a species is threatened. The national distribution of species is used by NCC as an essential measure of the conservation value of individual species and of assemblages (NCC 1989), but the range of taxonomic groups used in the SSSI selection criteria is limited to those for which comprehensive distribution information was in the public domain. The categorisation by NCC's Invertebrate Site Register (ISR) of invertebrate species as nationally rare or scarce (Ball 1986) has been based on information collated independently of BRC, although for some groups data at BRC were incorporated in the ISR.

Knowledge of the distribution of species provides a basis for assessing the status of species and, with the on-going surveys of national recording schemes, changes in status can be assessed over time. The data from such schemes provide a basis for future surveys and for monitoring (Rich 1992; Harding 1990b). Several recording schemes have moved on from what was an initial mapping-orientated phase to collect more detailed information on numbers of individuals and evidence of breeding; such information is essential for mobile species such as butterflies and dragonflies.

Site-relatable information in the database can be interrogated so that inventories of species occurring at sites can be compiled. The main difficulty with such an exercise is in defining the site itself for data retrieval. Geographical information systems allow the boundaries of sites to be held in digitised form and permit BRC records, referenced by geographical co-ordinates, to be related precisely to sites such as nature reserves and SSSIs. Trials in this procedure will be undertaken in collaboration with NCC in the near future.

The aggregation of information on species enables estimates of species richness to be made, at the level of individual sites, as a national overview for a taxonomic group (Figure 3) and as a national overview for a habitat assemblage (Figure 4). The selection of species can be varied to provide characteristic assemblages of species from a wide selection of taxonomic groups, eg the plants, birds and butterflies of wet heaths and moorland, or of ancient woodlands. The analysis of BRC data in these ways has only recently begun and has yet to be published.

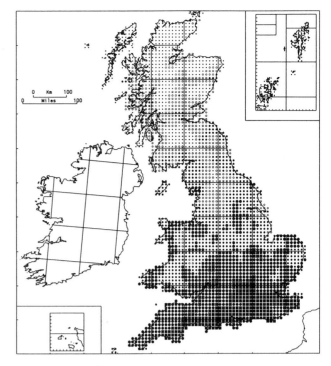

- ■ 36+ species
- ● 24–35 species
- • 15–23 species
- ○ 9–14 species
- · 1–8 species

Figure 3. Species richness map for butterflies in Britain. Data have been 'smoothed' over 30 x 30 km to reduce the patchiness caused by local variation in recorder effort

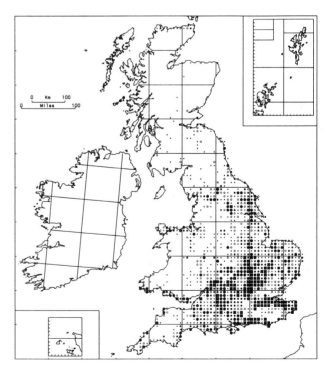

- ■ 5+ species
- ● 4 species
- • 3 species
- ○ 2 species
- · 1 species

Figure 4. Biotope assemblage mapping – calcareous grassland molluscs. A map of the recorded coincident occurrence (at 10 km square level) of a suite of species characteristic of calcareous grasslands

Published *Atlases* of distribution maps (see Harding 1989) provide a first line of reference for those concerned with site evaluation and with site and species conservation. Early BRC *Atlases* were merely collections of maps, with little or no commentary (eg Perring & Walters 1962; Heath 1970, 1978), but recent *Atlases* have provided detailed information on the habitats of species, on ecology and behaviour, and even on characteristic habitat assemblages (eg Heath *et al.* 1984; Harding & Sutton 1985; Hill *et al.* 1991).

Conservation policy is directed towards the preservation and management of semi-natural biotopes and their species assemblages. The extent to which such areas (SSSIs and nature reserves) already contain, and can continue to support, a representative suite of the species occurring in Britain is uncertain. NCC's SSSI selection criteria (NCC 1989) advocate the selection of sites based on factors such as the occurrence of nationally scarce or threatened species of a few taxonomic groups (vascular plants, vertebrates, butterflies, dragonflies and molluscs). Biological recording, using a large volunteer labour force, could be directed towards improving knowledge of these selected sites, but concentration of effort in such areas is unlikely to detect dynamic changes in the wider occurrence of species. Local changes, especially loss of species, are noticed and may be reported at a local level, but it is only when local information from the wider countryside, as well as from protected sites, is aggregated at a national level that the overall situation can be assessed. In this way, awareness of the serious decline in woodland fritillary butterflies was brought into stark perspective by the *Atlas* (Heath *et al.* 1984).

In these times of rapid environmental changes, particularly in land use and pollution, and, of course, the predicted changes in climate, there is an ever more pressing need to continue to update information on the occurrence of species and to enlarge the range of species covered in the national database at BRC. Only with comprehensive and up-to-date information on the occurrence of species in the wider countryside will conservation policy-makers be able to assess the effectiveness of past policies and to develop new policies for changing circumstances.

It is encouraging to note that NCC has recently taken an initial step in this direction by placing staff at BRC to collate information on 320 species of scarce vascular plants so that the status of these species can be assessed in relation to conservation priorities, such as inclusion in a revised *Red Data Book*.

CONCLUSIONS

From the original idea, developed some 40 years ago, to utilise volunteer specialists for collecting field information on the occurrence of species and to publish the results as species distribution maps, the Biological Records Centre has developed into a unique and unrivalled database on the occurrence of

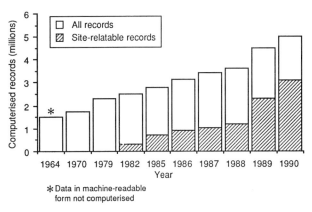

Figure 5. Biological Records Centre – growth of the computerised database

species in the British Isles. With its current data holdings of over 5 million records of more than 9000 species (Figure 5), and with on-going surveys covering some 16 000 species, BRC is a formidable data gathering and storage unit. The principles on which BRC has worked for 25 years have been a model for similar data centres elsewhere in Europe and in other continents.

The symbiotic relationship between professionals and amateurs which characterises the work of BRC is certainly unique within NERC and places the Council in the privileged position of collating data, through a spontaneous 'community programme', which can then be applied by the Council to its own research programmes and also be utilised in a variety of practical applications, such as nature conservation and environmental evaluation and monitoring.

It is almost certain that the existence of BRC, and the national recording schemes which it operates, have helped to fuel taxonomic research and ecological surveys by providing a focus for effort, by encouraging studies and the recruitment and training of new experts, and by providing a new realm of literature in the form of newsletters, *Atlases* and related publications. The milieu in which BRC operates extends far beyond its parent research council and NCC to include museums, national and local learned and natural history societies and specialist study groups, tertiary educational establishments, publishers, international agencies, and many more.

BRC has always had to walk the tightrope between the realities of its funding and resources, and the sometimes unrealistic, but not unreasonable, expectations of its volunteers. The precariousness of this balancing act has become increasingly acute in recent years as more schemes are reaching a stage when their data need to be processed, but when the resources available to BRC are, at best, static. Justifying the continuation of BRC in terms of the uses to which its data are put is essential, as is the ability to earn funding from commissions to supplement the steadily declining direct funding provided to NERC by the Government. It is no longer practical to consider BRC as part of 'the scientific equipment of the country' as was envisaged by Clapham and Godwin – a situation which will be familiar to anyone who has followed the recent fortunes of our national museums.

The application of the database has come to the fore in recent years – to provide material for research publications, to provide site information for nature conservation and, of course, to provide new and updated information on species distributions to a wide variety of users, including the volunteers. Within ITE's new Environmental Information Centre, BRC will have a key role in providing spatially referenced data on species and assemblages. Although the requirements of the four bodies which succeeded the Nature Conservancy Council in April 1991 are as yet not clearly defined, it is to be expected that BRC will have a role in contributing to their information resources. General environmental awareness and statutory requirements for environmental assessments in relation to major development proposals are already leading to an increasing demand for information from BRC.

The challenge for the future will be to strike a balance between utilising the existing information, and acquiring new information to update and enlarge the database. New computer-based technologies must be exploited to the full so that the scarce resources of manpower can be used to best effect. In reviewing the last 25 years of BRC, we have been struck by the very real influences of technological advances, political decisions, and personalities on the work of the Centre. There is no reason to believe that subsequent years will not experience similar influences!

ACKNOWLEDGEMENTS

We are grateful to the Nature Conservancy Council for granting access to the relevant files of the former Nature Conservancy, and to Dr B K Wyatt for helpful comments on a draft of this paper.

REFERENCES

Anon. 1985. Biological recording and the use of site-based biological information. *Biology Curators' Group Newsletter,* **4**, Supplement, 1–72.

Ball, S.G. 1986. *Terrestrial and freshwater invertebrates with Red Data Book, Notable or Habitat Indicator status.* (CSD report no. 637.) Peterborough: Nature Conservancy Council.

Berry, R.J. 1988. *Biological survey: need and network.* London: Linnean Society of London/PNL Press.

Bunce, R.G.H., Barr, C.J. & Whittaker, H.A. 1981. An integrated system of land classification. *Annual Report of the Institute of Terrestrial Ecology 1980,* 28–33.

Clapham, A.R. 1951. A proposal for mapping the distribution of British plants. In: *The study of the distribution of British plants,* edited by J.E. Lousley, 110–117. Oxford: Botanical Society of the British Isles.

Copp, C.J.T. & Harding, P.T. 1985. *Biological Recording Forum 1985.* (Biology Curators' Group special report no. 4.) Bolton: Biology Curators' Group.

Council of Europe. 1985. *Catalogue of data banks in the field of nature conservation.* (CDSN-INF (85) 2.) Strasbourg: Council of Europe.

Council of Europe. 1990. *Colloquy on the Berne Convention invertebrates and their conservation: conclusions and summaries.* (Environmental encounters series, no. 10.) Strasbourg: Council of Europe.

Ellis, R.G. 1986. Botanical Society of the British Isles monitoring scheme. In: *Biological recording in a changing landscape,* edited by P.T. Harding & D.A. Roberts, 6–9. Cambridge: National Federation for Biological Recording.

Flood, S.W. & Perring, F.H. 1978. *A handbook for local biological records centres.* Huntingdon: Institute of Terrestrial Ecology.

Hammond, C.O. 1977. *The dragonflies of Great Britain and Ireland.* London: Curwen.

Harding, P.T. 1989. *Current atlases of the flora and fauna of the British Isles.* Huntingdon: Biological Records Centre.

Harding, P.T. 1990a. Famous laboratories: the Biological Records Centre. *Biologist,* **37**, 162–164.

Harding, P.T. 1990b. National species distribution surveys. In: *Monitoring for conservation and ecology,* edited by F.B. Goldsmith, 133–154. London: Chapman and Hall.

Harding, P.T. 1990c. Biological checklists, a European perspective. In: *Terminology for museums,* edited by D.A. Roberts, 441–446. Cambridge: Museum Documentation Association.

Harding, P.T. 1990d. Biological survey: need and network. A review of progress towards national policies. In: *National perspectives in biological recording in the UK,* edited by G. Stansfield & P.T. Harding, 1–15. Cambridge: National Federation for Biological Recording.

Harding, P.T. & Greene, D.M. 1984. Butterflies in the British Isles: a new data base. *Annual Report of the Institute of Terrestrial Ecology 1983,* 18–19.

Harding, P.T. & Greenwood, E. 1981. Survey of local and regional biological records centres – inventory. *Biology Curators' Group Newsletter,* **2**, 468–478.

Harding, P.T. & Sutton, S.L. 1985. *Woodlice in Britain and Ireland: distribution and habitat.* Huntingdon: Institute of Terrestrial Ecology.

Harding, P.T., Rushton, S.P., Eyre, M.D. & Sutton, S.L. 1991. Multivariate analysis of British data on the distribution and ecology of terrestrial Isopoda. In: *Biology of terrestrial isopods,* III, edited by P. Juchault & J. P. Mocquard, 65–72. Poitiers: Université de Poitiers.

Hawksworth, D.L. & Seaward, M.R.D. 1990. Twenty-five years of lichen mapping in Great Britain and Ireland. *Stuttgarter Bieträge zur Naturkunde, Series A,* **456**, 5–10.

Heath, J. 1970. *Provisional atlas of the insects of the British Isles; Part 1, Lepidoptera Rhopalocera – butterflies.* Huntingdon: Biological Records Centre.

Heath, J. 1971. Insect distribution maps scheme progress report 1971. *Entomologist,* **104**, 305–310.

Heath, J. 1975. *European Invertebrate Survey. Proceedings of the 2nd International Symposium, 1973.* Huntingdon: Institute of Terrestrial Ecology.

Heath, J. 1978. *Provisional atlas of the insects of the British isles; Part 7, Odonata – dragonflies.* Huntingdon: Biological Records Centre.

Heath, J. & Skelton, M.J. 1973. *Provisional atlas of the insects of the British Isles: Part 2, Lepidoptera (Moths – part one).* Huntingdon: Biological Records Centre.

Heath, J., Pollard, E. & Thomas, J. 1984. *Atlas of butterflies in Britain and Ireland.* Harmondsworth: Viking.

Hill, M.O., Preston, C.D. & Smith, A.J.E. 1991. *Atlas of bryophytes of Britain and Ireland: 1, Liverworts (Hepaticae and Anthocerotae).* Colchester: Harley Books.

Howarth, T.G. 1973. *South's British butterflies.* London: Warne.

Hultén, E. 1950. *Atlas of the distribution of vascular plants in northwest Europe.* Stockholm: Generalstabens Litografiska Anstalts.

Ing, B. 1982. *Provisional atlas of the myxomycetes of the British Isles.* Huntingdon: Biological Records Centre.

Jalas, J. & Suominen, J., eds. 1972. *Atlas Florae Europaeae, Vol. 1, Pteridophyta (Psilotaceae to Azollaceae).* Helsinki: Committee for Mapping the Flora of Europe and Societas Biologica Fennica Vanamo.

Jalas, J. & Suominen, J., eds. 1973. *Atlas Florae Europaeae, Vol. 2, Gymnospermae (Pinaceae to Ephidraceae).* Helsinki: Committee for Mapping the Flora of Europe and Societas Biologica Fennica Vanamo.

Jermy, A.C., Chater, A.O. & David, R.W. 1982. *Sedges of the British Isles.* (BSBI Handbook no. 1.) 2nd ed. London: Botanical Society of the British Isles.

Luff, M.L., Rushton, S.P., Eyre, M.D. & Foster, G.N. 1992. Environmental and ecological applications of invertebrate distribution data from the Biological Records Centre. In: *Biological recording of changes in British wildlife,* edited by P.T. Harding, 74–75. London: HMSO.

Merritt, R., Moore, N.W. & Eversham, B.C. 1992. *Atlas of the dragonflies of Britain and Ireland.* London: HMSO. In press.

Moore, J.A. 1986. *Charophytes of Great Britain and Ireland.* (BSBI Handbook no. 5.) London: Botanical Society of the British Isles.

Moore, J.A. & Greene, D.M. 1983. *Provisional atlas and catalogue of British Museum (Natural History) specimens of the Characeae.* Huntingdon: Biological Records Centre.

Moss, D. & Wyatt, B.K. 1990. *CORINE biotopes project: an inventory of sites of importance for nature conservation in the European Community.* (Final Report to the Commission of the European Communities on Contracts 6601(88)06, 6601(89)02 and 6601(90)08.) Huntingdon: Institute of Terrestrial Ecology.

Natural Environment Research Council. 1973. *Report for the year 1972–1973.* London: HMSO.

Nature Conservancy. 1962. Botanical Society's atlas. In: *Report of the Nature Conservancy for the year ended 30th September 1962,* 44–45. London: HMSO.

Nature Conservancy. 1964. Biological Records Centre. In: *Report of the Nature Conservancy for the year ended 30th September 1964.* 53–54. London: HMSO.

Nature Conservancy Council. 1989. *Guidelines for the selection of biological SSSIs.* Peterborough: Nature Conservancy Council.

Norton, T.A. 1985. *Provisional atlas of the marine algae of the British Isles.* Huntingdon: Institute of Terrestrial Ecology.

Perring, F.H. 1967. Mapping the flora of Europe. *Proceedings of the Botanical Society of the British Isles,* **6**, 354–357.

Perring, F.H. 1970. The last seventy years. In: *The flora of a changing Britain,* edited by F.H. Perring, 128–135. Hampton: Classey.

Perring, F.H. 1971. Rare plant recording and conservation in Great Britain. *Boissera,* **19**, 73–79.

Perring, F.H. & Farrell, L. 1977. *British Red Data Books: 1, Vascular plants.* Lincoln: Royal Society for Nature Conservation.

Perring, F.H. & Farrell, L. 1983. *British Red Data Books: 1, Vascular Plants.* 2nd ed. Lincoln: Royal Society for Nature Conservation.

Perring, F.H. & Sell, P.D. 1968. *Critical supplement to the atlas of the British flora.* London: Nelson.

Perring, F.H. & Walters, S.M. 1962. *Atlas of the British flora.* London: Nelson.

Pollard, E. 1992. Monitoring populations of a butterfly during a period of range expansion. In: *Biological recording of changes in British wildlife,* edited by P.T. Harding, 60–64. London: HMSO.

Pollard, E., Hall, M.L. & Bibby, T.J. 1986. *Monitoring the abundance of butterflies, 1976–1985.* (Research and Survey in Nature Conservation no. 2.) Peterborough: Nature Conservancy Council.

Preston, C.D. 1990. An index and bibliography to distribution maps published between 1962 and 1989. In: *Atlas of the British flora,* edited by F.H. Perring & S.M. Walters (3rd ed. reprinted), 426–434. London: Botanical Society of the British Isles.

Rich, T.C.G. 1992. BSBI monitoring scheme (1987–88). In: *Biological recording of changes in British wildlife,* edited by P.T. Harding, 75. London: HMSO.

Seaward, M.R.D. & Hitch, C.J.B. 1982. *Atlas of the lichens of the British Isles: Vol. 1.* Cambridge: Institute of Terrestrial Ecology.

Sheail, J. 1987. *Seventy-five years in ecology: the British Ecological Society.* Oxford: Blackwell Scientific.

Smith, A.J.E. 1978. *Provisional atlas of the bryophytes of the British Isles.* Huntingdon: Biological Records Centre.

Sommerville, A. 1977. *A guide to biological recording in Scotland.* Edinburgh: Scottish Wildlife Trust for the Biological Recording in Scotland Committee.

Stansfield, G., ed. 1973. *Centres for environmental records.* (Vaughan papers in adult education, no. 18.) Leicester: University of Leicester, Department of Adult Education.

Sutton, S.L. & Harding, P.T. 1989. Interpretation of the distribution of terrestrial isopods in the British Isles. *Monitore zoologico italiano (N.S.) Monografia,* **4**, 43–61.

Taylor, E.G.R. 1940. Plans for a national atlas. *Geographical Journal,* **95**, 96–108.

Tutin, T.G., Heywood, V.H., Burges, N.A., Valentine, D.H., Walters, S.M. & Webb, D.A. 1964. *Flora Europaea: Vol. 1, Lycopodiaceae to Plantanaceae.* Cambridge: Cambridge University Press.

Wyatt, B.K. 1992. Resources for documenting changes in species and habitats. In: *Biological recording of changes in British wildlife,* edited by P.T. Harding, 20–26. London: HMSO.

Yates, T.J. 1992. The butterfly monitoring scheme. In: *Biological recording of changes in British wildlife,* edited by P.T. Harding, 77–78. London: HMSO.

MANUSCRIPT REFERENCES

Superscript numbers in the text refer to the following manuscript sources which are held by English Nature (1–4) and by the Biological Records Centre (5–12).

1 Nature Conservancy Council (NCC), NC, 0 34.

2 NCC, NC, G/M/61/3 & F 215, vol. 2.

3 NCC, NC, F 215, vol. 2.

4 NCC, NC, Min 63/3, item 14; F 215, vol. 3.

5 BRC Advisory Sub-Committee minutes (ASC), 12/63 & 5/64.

6 ASC, 10/66.

7 ASC, 5/66.

8 ASC, 10/67.

9 ASC, 5/68.

10 ASC, 10/64, 5/65 & 10/65.

11 ASC, 5/68 & 10/69.

12 ASC, 5/64 & 5/68.

Resources for documenting changes in species and habitats

B K Wyatt

Environmental Information Centre, Institute of Terrestrial Ecology, Monks Wood Experimental Station, Abbots Ripton, Huntingdon, Cambs PE17 2LS

INTRODUCTION

It is fashionable to regard environmental change as if change were an abnormal condition. In fact, the converse is true. Stability is the exception rather than the rule in the natural environment and change is a normal condition in ecological systems. Indeed, environmental change is the driving force for evolution. It is particularly interesting to speculate whether man, despite his present anxiety over the pace of environmental change, will succeed in becoming the first species to stem the tide of evolution through the degree of control he can exert over his environment.

The history of organised observation of the natural world has demonstrated a recent dramatic increase in the rate, the extent and the amplitude of change, and much of this increase has been directly or indirectly the result of man's activities.

It is hardly necessary in this paper to list the individual causes of environmental change. By now, we have become all too familiar with the consequences of man's success as a species and of the increasing demands his growing population places on the natural resources of this planet.

The effects of such changes on biological species may be direct – destruction of habitat or environmental pollution, for example, or indirect – the fragmentation of habitats, preventing the free movement of species, or the elimination of important components in foodchains. Recently we have become aware of the danger of more widespread changes in environmental conditions, notably the risk of global climate change resulting from increases in the concentration of 'greenhouse gases'.

In practice, our response to such threats usually takes the form of a compromise between our demand for material advancement and the need to preserve the natural environment which we all value and upon which we all ultimately depend. This compromise is necessary because, despite the undoubted attractions of the pre-industrial environment, few of us would be prepared to return to the social conditions of that period. Therefore, the only strategies for conservation and protection which carry with them a real prospect of success are those based on the principles of sustainable development (Pearce, Barbier & Markardya 1990).

To achieve a satisfactory compromise, we need information on conditions and trends in the various physical and biological components of the biosphere. We need information on baseline conditions, we need reliable indicators of change, and we need an understanding of how ecosystems respond to change, so that we may develop predictive models of these responses and take the required preventive action. The information needed often relates to long-term trends within a pattern of cyclic variation and random 'noise'. The long timescales involved, and the frequently poor signal-to-noise ratio of our monitoring systems, require considerable ingenuity in the manner in which we make use of available data.

Yet, ironically, the volume of information available to us and our skill in handling it provide instances where we can point to significant advances in our ability to respond to environmental threats. Never before has so much been known about current environmental conditions and about the way in which organisms and ecosystems respond to change.

In the remainder of this paper, I shall discuss, through examples, some of the ways in which information technology can contribute to environmental science and to environmental protection. The paper is intended as a bridge between the historical perspectives of the earlier papers and the following papers which examine how we are beginning to harness the growing power and sophistication of information processing systems to improve our understanding and capacity for effective environmental management.

BIOLOGICAL RECORDING

Before discussing the environmental databases available to us, some definition of 'biological recording' is needed. The Linnean Society, host for this Conference, was instrumental in the formation of a recent Working Party to consider how best to establish an effective network for biological surveillance.

This Working Party defined biological recording as:

> '. . . the collection, collation, storage, dissemination and interpretation of spatially and temporally referenced information on the occurrence of biological taxa, assemblages and biotopes. Basic information on occurrence is normally augmented and amplified with a variety of related

biological, environmental and administrative information. Biological recording normally excludes information on agricultural, horticultural or forestry crops, and agricultural, domestic or captive stock, except where it may concern wildlife, biotopes or the management of semi-natural areas.' (Co-ordinating Commission for Biological Recording 1989)

I shall use this definition in the subsequent discussion.

The preceding paper (Harding & Sheail 1992) described the Biological Records Centre (BRC) – the national focus of the network of biological recording in Great Britain. In this paper, I propose to review some of the other elements in this network, which help to ensure the continuity of the biological surveillance on which BRC largely depends to maintain the national archive.

Figure 1 provides an overview of recent (post-1980) developments in the evolution of this network, described in greater detail by Harding (1990).

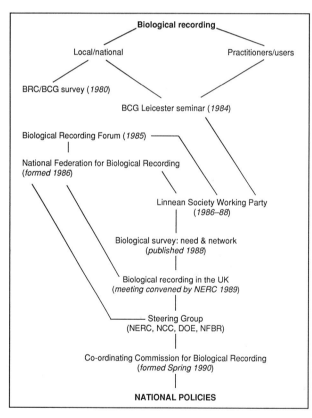

Figure 1. Progress towards national policies in biological recording (after Harding 1990)

LOCAL RECORDING

Although locally based collection of information on wildlife has been in evidence for at least 150 years, formalised environmental recording at the local level only came to the fore in the early 1970s. A conference in 1973 brought together many of those concerned with local biological records centres (Stansfield 1973) and, in 1977, the Museums Association convened a standing committee on

Environmental Records Centres (Stewart 1980). Also in 1977, the Biological Recording in Scotland Committee (BRISC) was formed (Sommerville 1977). Following a meeting of records centre organisers at Monks Wood in December 1977, a *Handbook for local biological records centres* was published in 1978 (Flood & Perring 1978).

BIOLOGY CURATORS' GROUP

In the absence of any other co-ordinating group, many local centre managers looked to the Biology Curators' Group (BCG). BCG convened an important seminar in Leicester in 1984 to discuss biological recording. The seminar concluded:

1. that existing arrangements for recording, storing and retrieval of biological data were unsatisfactory and under-funded;

2. that agreed standards for biological recording should be set, bearing in mind the needs of both amateur naturalist and professional scientist;

3. that museums should, where possible, manage local biological databanks which should provide a range of information services to the general public.

The seminar resulted in two initiatives. At a practical level, in 1985, BCG and BRC jointly set up the Biological Recording Forum (Copp & Harding 1985), from which the National Federation for Biological Recording (NFBR) evolved in 1986. On the political front, the Linnean Society agreed to convene, under the chairmanship of Professor S Berry, the Working Party on Biological Surveillance, which brought to public awareness many of the issues needing to be addressed to establish biological recording on a sounder basis for the future.

Independently of these two initiatives, in 1986 the Nature Conservancy Council (NCC) and Wildlife Link formed the Joint NCC/NGO Data Handling Group.

Concurrent with all this activity (and perhaps a stimulus for some of it), there have been significant changes in both the purposes and the methods of biological recording. Computer-based methods of information management are increasingly in evidence and introduce strong pressures for rationalisation and harmonisation of data structures. The implementation of the RECORDER package for the management of computerised data by local records centres and wildlife trusts, through collaboration between NCC, the Royal Society for Nature Conservation, and the World Wide Fund for Nature, will inevitably accelerate this trend.

In the early years, mapping the distribution of species was almost the sole justification for recording. More recently public concern for the quality of the environment has led to increasing demands from Government and developers for biological information at all stages of the planning process, and especially in drawing up environmental impact statements. The emphasis has therefore shifted rapidly towards collecting data which relate species records to individual sites and thus to habitat.

These demands have opened up new markets for information at both local and national levels; however, the trend can result in a conflict of interest between the data provider and the custodian of the data; neither is it likely that commercial revenue alone will be sufficient to provide the necessary investment in capital and staff to ensure the efficient operation of a national network for biological recording.

THE LINNEAN SOCIETY REPORT

In 1986, the Linnean Society's Working Party on Biological Surveillance addressed these problems. Following publication of its report (Berry 1988), a meeting of more than 30 national organisations with an interest in biological surveillance was held at the Royal Society, under the chairmanship of Dr P B Tinker, Director of Terrestrial and Freshwater Sciences in the Natural Environment Research Council (NERC).

The meeting endorsed most of the recommendations of the Linnean Society report, and set up a steering group to suggest terms of reference and membership of a Co-ordinating Commission for Biological Recording. Once established, this Co-ordinating Commission would establish procedures for accreditation of participating groups and would oversee the operation of a national network.

The steering group reported at the end of 1989; Sir John Burnett has since accepted the invitation to convene and chair the Co-ordinating Commission itself, which has now prepared a statement of intent and a programme for the establishment of a national system (Co-ordinating Commission for Biological Recording 1990). The Commission is a unique development, in that it brings together those with an interest in using biological data and the recording community. It is likely to have a profound and beneficial impact on biological recording in the future.

ENVIRONMENTAL INFORMATION CENTRE

At the same time as these developments in the organisation of biological recording, NERC announced the formation of the Environmental Information Centre (EIC) at the Institute of Terrestrial Ecology (ITE), Monks Wood, to serve as its data centre in the terrestrial life sciences (Figure 2). The objectives of EIC are to develop improved methods for the storage, processing and analysis of ecological data, to enhance the ability to inter-relate datasets describing different aspects of the natural environment, and to improve the relevance of these information systems for applications in ecological research, in planning, and in environmental protection.

EIC brings together many of the groups in ITE with expertise in large and complex biological databases,

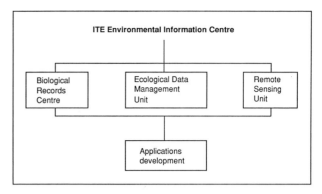

Figure 2. The Environmental Information Centre of ITE, a data centre for terrestrial ecology and rural land use – schematic structure

and particularly in the handling of geographically referenced data. The Biological Records Centre forms the single largest unit in EIC, with the most extensive data resource. Elsewhere in EIC, specialists are researching the use of data from earth observation satellites to map and to monitor changes in the land surface. Modern computer-based geographical information systems provide a powerful means of exploring relationships between environmental conditions and species distributions. EIC is beginning to apply these systems to the BRC database and to many other environmental datasets, to look at national trends in relation to soils, climate and environmental quality, and also to study ecosystems more locally at specific sites.

EIC currently offers access to digital information on species and habitat and to digital topographic and thematic maps covering environmental variables such as soil type, climate and land use. The Centre has been in existence for less than two years, but evidence of its practical benefits is already apparent: in improved access to information and expertise; in the integration of ecological data; in technology transfer within the Institute; and in the links to other disciplines. The growing interest in research into global environmental change will place a premium on many of these qualities – in particular, on the ability to access multidisciplinary data and expertise in pursuit of the understanding of complex, large-scale ecological processes.

A major development over the next two years will be the compilation, from remote sensing, of a national digital map of land cover. Information on the present and future disposition of natural and managed habitats is crucial to our understanding of how the environment will respond to future changes. Remote sensing will allow us to map the present situation, to generate a baseline against which to measure change, and subsequently to monitor these changes and their consequences. The land cover map is being compiled as part of the Countryside Survey 1990, a joint venture between NERC, the Department of the Environment and the British National Space Centre. The Countryside Survey includes intensive ground survey within sample areas, and this ground-based survey will generate reference data with

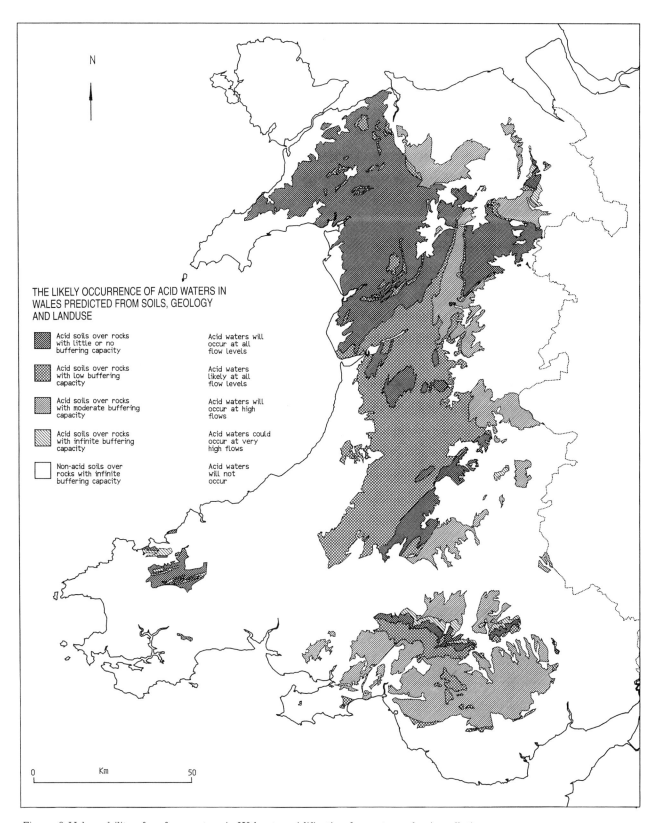

THE LIKELY OCCURRENCE OF ACID WATERS IN
WALES PREDICTED FROM SOILS, GEOLOGY
AND LANDUSE

Acid soils over rocks
with little or no
buffering capacity

Acid waters will
occur at all
flow levels

Acid soils over rocks
with low buffering
capacity

Acid waters
likely at all
flow levels

Acid soils over rocks
with moderate buffering
capacity

Acid waters will
occur at high
flows

Acid soils over rocks
with infinite buffering
capacity

Acid waters could
occur at very
high flows

Non-acid soils over
rocks with infinite
buffering capacity

Acid waters
will not
occur

Figure 3. Vulnerability of surface waters in Wales to acidification from atmospheric pollution

which to calibrate the satellite maps, as well as more detailed records of field conditions than it is possible to observe from space.

Geographical information systems (GIS) allow us to explore spatial and temporal relationships between environmental variables, using spatial overlay and similar techniques. For example, information on soils in north Wales has been overlain with information on land use and with geological data to generate a map showing areas where surface waters are particularly vulnerable to acidification from atmospheric pollution (Figure 3). GIS allow us to extract information on surface water quality from an associated database in order to calibrate the map and to use the data predictively.

An extremely powerful technique is to use the GIS to process information from remote sensing in combination with digital map data. EIC is particularly well-equipped to develop and exploit these approaches, and a number of pilot-scale projects have been undertaken to demonstrate their potential for land resource planning, for environmental assessment, and for ecological research. In one example (Jones & Wyatt 1989), Landsat Thematic Mapper imagery was integrated with topographic map data, with a digital elevation model, and with thematic maps recording information on soil type, geology and hydrology for the whole of the Snowdonia National Park in north Wales. The resulting geographical database was used to evaluate the impacts on ecology and landscape of various proposed economic developments in the National Park, including the effects of forestry, tourism, and industry.

A second example uses remotely sensed imagery to delineate areas of mudflats used as feeding grounds by wading birds. Counts of bird numbers, measured in the field and held in a digital database, have been related to characteristics of the mudflats, which can be distinguished in the satellite imagery. Bird populations are correlated with sediment type because food supplies vary with, for example, sediment grain size or wetness. The calibration of bird numbers with mudflat type can then be applied over much more extensive areas in order to allow estimation of bird populations with less effort and at greater precision than has previously been possible.

These techniques have been applied at a variety of spatial scales and at different levels of generalisation. Some of the most ambitious applications have been in a European context, through the development of CORINE (Co-ordinated Environmental Information in the European Community) – an experimental programme of the Environment Directorate of the European Commission. The CORINE programme is generating a geographically referenced database covering the entire territories of the European Community and recording more than 50 different environmental variables from every sector of the environment. CORINE is intended to lead to improved policy decisions on the environment in the Community, and to help the Commission to assess the consequences of its development proposals for environmental resources.

EIC has provided technical experts who have assisted the European Commission in the design and development of the CORINE database, which has now been in operation for five years, providing an effective illustration of the potential of GIS for environmental planning and management. The principal contribution of EIC has been to design and implement a system for recording information on areas of importance for nature conservation on a Community-wide basis. This information has been used, for example, to identify important sites threatened by major development projects (Figure 4), or to help in drawing up the draft Community environmental protection legislation by examining the extent of threatened species or habitats in the Community as a whole (Figure 5).

We are beginning to use similar methods for analysing the national environmental databases which we are building, to predict and measure the ecological consequences of future changes in land use or environmental conditions.

SUMMARY

The use of computers for the collection, storage, management and analysis of environmental data has enormous benefits, in the improved efficiency with which existing information can be handled, in the increased volumes and complexity of the information that such systems can hold, in the relative ease with which data can be exchanged, and in the facility with which sophisticated analysis can be undertaken.

In this paper, I have illustrated examples of ways in which some of these benefits can be realised, particularly in relation to the computer-based analysis of spatially referenced data to model the consequences of environmental change.

However, these benefits are not totally without cost; if computer systems are to be exploited to their full potential, then it is necessary that the providers of data exercise discipline and adhere to basic minimum standards with regard to the form, the consistency and the accuracy of the data and that the users of such systems observe any restrictions which may be necessary, either because of characteristics of the data (eg precision) which constrain their use or because of other restrictions on their dissemination (eg considerations of confidentiality).

Such restrictions can be applied relatively easily in an environment where a single organisation is responsible for all aspects of the database. Biological recording rarely conforms to this pattern; by its nature, biological surveillance tends to involve many different subject specialisms, from many different organisational backgrounds. Comprehensive biological recording over large areas (eg nationally) can only be achieved through loose affiliations of distributed organisations and individuals. In this environment, enforcement is, at best, extremely diffi-

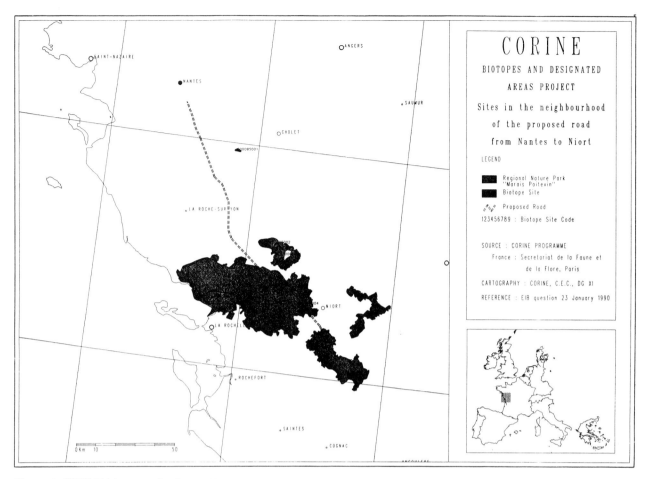

Figure 4. CORINE biotopes database. Biotopes and designated areas in the neighbourhood of the proposed road from Nantes to Niort

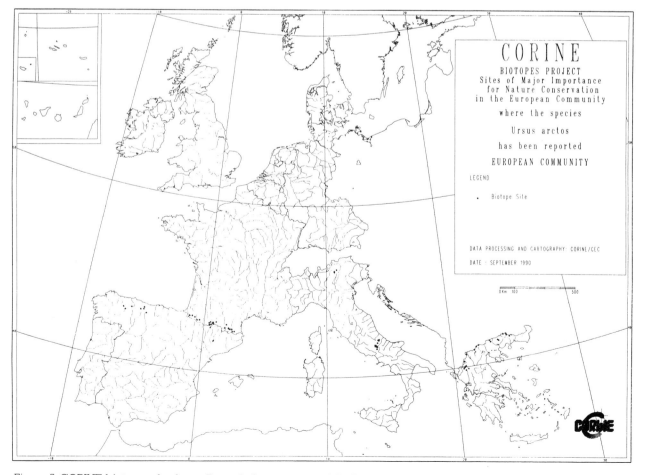

Figure 5. CORINE biotopes database. Recorded occurrence of the brown bear (*Ursus arctos*) in the European Community

cult. The solution most likely to succeed will depend on enlightened self-interest; each participant will recognise the advantages of adopting consistent standards. The work of the Co-ordinating Commission for Biological Recording, in drawing up the necessary standards, in ensuring their widespread adoption and in providing the necessary software support and documentation, will clearly be of crucial importance.

REFERENCES

Berry, R.J. 1988. *Biological survey: need and network.* London: Linnean Society of London/PNL Press.

Co-ordinating Commission for Biological Recording. 1989. *Report of the steering group set up at the meeting 'Biological recording in the UK'.* Huntingdon: Institute of Terrestrial Ecology. (Unpublished.)

Co-ordinating Commission for Biological Recording. 1990. *Programme for the establishment of a national system for the co-ordination of biological recording.* Huntingdon: Institute of Terrestrial Ecology. (Unpublished.)

Copp, C.J.T. & Harding, P.T., eds. 1985. *Biological Recording Forum 1985.* (Biology Curators' Group Special Report no. 4.) Bolton: Biology Curators' Group.

Flood, S.W. & Perring, F.H. 1978. *A handbook for local biological records centres.* Huntingdon: Institute of Terrestrial Ecology.

Harding, P.T. 1990. Biological survey: need and network. A review of progress towards national policies. In: *National perspectives in biological recording in the UK,* edited by G. Stansfield & P.T. Harding, 1–15. Cambridge: National Federation for Biological Recording.

Harding, P.T. & Sheail, J. 1992. The Biological Records Centre: a pioneer in data gathering and retrieval. In: *Biological recording of changes in British wildlife,* edited by P.T. Harding, 5–19. London: HMSO.

Jones, A.R. & Wyatt, B.K. 1989. Remote sensing for monitoring and inventory of protected landscapes: resource management in a less favoured area. In: *Remote sensing for operational applications,* compiled by E.C. Barrett & K.A. Brown, 193–199. Nottingham: Remote Sensing Society.

Pearce, D., Barbier, E. & Markardya, A. 1990. *Sustainable development: economics and environment in the Third World.* London: E. Elgar Publications. (Reprinted 1991. London: Earthscan).

Sommerville, A. 1977. *A guide to biological recording in Scotland.* Edinburgh: Scottish Wildlife Trust for the Biological Recording in Scotland Committee.

Stansfield, G., ed. 1973. *Centres for environmental records.* (Vaughan papers in adult education, no. 18.) Leicester: University of Leicester.

Stewart, J.D. 1980. A summary of local environmental record centres in Britain. *Museums Journal,* **80**, 161–165.

The effects of environmental changes on wildlife

The effects of changes in land use on water beetles

G N Foster

Scottish Agricultural College, Auchincruive, Ayr KA6 5HW; and The Balfour-Browne Club, 3 Eglinton Terrace, Ayr KA7 1JJ

INTRODUCTION

The aquatic Coleoptera recording scheme was formally initiated in July 1979 as an activity of the Balfour-Browne Club. The present field record card (Biological Records Centre RA71) lists 353 species in 15 families and signifies the unusual feature of the scheme in that it is based more on an ecological grouping than on a systematic one. The database has yet to be compiled on a mainframe computer. It is estimated that there are at least 80 000 species records for 10 km squares and 5000 site listings that can be classed as 'complete' in the sense that the recorder made an attempt to record all taxa at a site, rather than just the rare species or those of a particular family.

The scheme was developed on traditional lines, with emphasis on recording of individual species for the express purpose of mapping the 10 km distribution of each species. A major input to the database was derived from the vice-county recording scheme of Professor Frank Balfour-Browne. Coverage of most of Britain is now good and resembles that of many other schemes, with patchy cover in northern Scotland and parts of Wales. Modern coverage of Ireland is poor, recent discoveries of relict species there underlining the need for more recording effort (Bilton 1988).

The habitat basis of recording, rather than a taxonomic grouping, has led to increasing emphasis on the use to which site lists can be put in the cause of conservation. This shift in emphasis, from amateur recording of individual species to 'professional' evaluation of site data, has thus far resulted in few conflicts because it is still possible to service the primary interest in species maps. However, one can never escape the time-honoured criticisms of national recording schemes – the uncritical acceptance of a dubious record, the failure to register a particular record, the failure to place records in the correct grid square, and, most of all, the possibility that species maps reflect the distribution of recording activity rather than the true distribution of the species. The new approach simply adds a new range of criticisms.

CHANGES IN FAUNA

Changes are in three main categories:

- species extinctions and additions;
- changes in distributions of species;
- changes in species assemblages.

The British checklist

The crudest estimates of change are the gains and losses to the British checklist. Since the completion of Professor Balfour-Browne's *British water beetles* (1940, 1950, 1958), 20 species have been added to the British water beetle list. Of these, four are dubious, being based on old material in Continental museums and 16 are the result of revisional work and redefinition of species, mainly hydrophilids. Of the nine species that have gone unrecognised in Britain until the 1970s and 1980s, at least four are true relict species and four are pioneer species first discovered in the Weald, two of them being confined to recently created ponds. To these nine may be added five 'non-extinctions', species claimed to be extinct, but later rediscovered. Against these gains must be set three losses based on spurious records from the Hebrides, six species that truly appear to have become extinct, mainly being found last at the turn of the century, and three that are known only as post-glacial, sub-fossil fragments.

This turnover in the British checklist produces the highlights when proclaiming progress in the recording scheme, but it is often associated with name changes, the need for which is not understood by many ecologists and amateur entomologists. A dynamic checklist attracts species hunters, but can be detrimental to ecological studies; ecological work is also hampered by an incomplete knowledge of the immature stages of water beetles.

Changes in species distribution

With about 300 species maps from which to choose, it should be possible to find at least one to demonstrate any particular change in land use. This *a posteriori* approach is dangerous, but is of great value in generating speculation, discussion and more data. The main change demonstrated by species maps is habitat loss. Rather than choose an obvious example, I have chosen *Hydroporus striola*, a typical

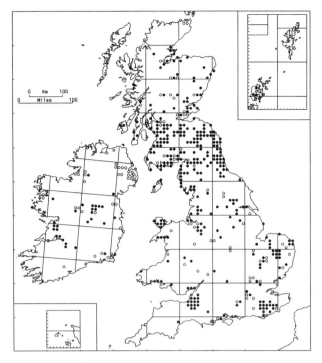

● 1950 onwards
○ pre-1950

Figure 1. Known distribution of *Hydroporus striola* (Col., Dytiscidae) in Britain and Ireland in 1990

dytiscid pond and marsh species, to illustrate some of the problems with which amateur recorders must contend. The distribution (Figure 1) is largely lowland, with major gaps in Ireland easily explained by lack of recording activity. The lack of records from south-west England must at least indicate that the species is very rare there. The species is recorded from most fen systems, but, unlike several 'fen' species, it is rare in Essex and in the arable fenland drainage system south of the Wash. It can be assumed that this rareness resulted from drainage of the East Anglian fens, by which stagnant habitats were replaced by drains, since *H. striola* rarely occurs in flowing water. A larger area of sparse records is associated with the tract of midland and southern England referred to as 'planned country-side' by Rackham (1986), the result of open field agriculture developed following the Enclosure Acts of the 18th century. In contrast, *H. striola* is densely recorded from the 'ancient countryside' of areas such as south-east England and Cheshire. By counting ponds on the 1:25 000 Ordnance Survey maps, Rackham (1986) showed that the planned country-side of lowland England had about five ponds per square mile, whereas there were 12 per square mile in ancient countryside. Some areas of Norfolk, Suffolk and Cheshire had 30 ponds per square mile, and this number was 'vastly exceeded by the ponds of some ancient woods'. It might thus be claimed that '*H. striola* was once abundant in most of lowland Britain, except on freely draining chalk and limestone, but never reached south-west England, the Outer Hebrides or the Shetlands. Distribution was

later restricted by drainage of stagnant habitats in the Fens and by infilling of ponds in the 18th and mid-20th centuries.'

An alternative explanation might run: 'The distribution reflects the intensity of recording of an obscure *Hydroporus* species associated with the main areas worked by a handful of enthusiasts, in particular one who was at various times based in East Sussex, Cambridgeshire, Anglesey, Northumberland and Ayrshire'.

It is not possible to reconcile the views of the enthusiast and the cynic, except by reference to a complete post-glacial record of distributional changes – either by use of a time machine, or by reference to a massive study of sub-fossil material in post-glacial deposits.

Land use changes

The main post-glacial changes encountered by water beetles (and most other organisms) in the British Isles are summarised in Table 1. It is possible to claim an extreme case associated with each change, but it is more important to emphasise that habitat changes and losses are never uniform. Thus, the general deterioration in climate from the Mesolithic period onwards has resulted in apparent southern distributions with northern refugia, whereas agricultural improvement in the planned countryside, industrial development in the midlands and urbanisation have brought about apparent northern distributions with southern refugia.

The southern species with northern refugia must be distinguished from those thermophilous species that made frequent sorties into northern Britain. Their 'pre-1950, post-1950' maps look the same. Here a careful study of the habitats occupied in northern Britain will differentiate the relicts from the opportunists. Unfortunately, relict sites are often the first in which newly invading opportunistic species are detected, because such sites are frequently visited by recorders.

Drainage and resultant habitat loss are clearly the most important features of land use changes. However, some species have benefitted from change. For example, the dytiscid *Agabus melanarius* is largely restricted to relict forest in mainland Europe, but it has become established in older conifer plantations in northern and south-west England (Barneul, Foster & Holmen 1982). It is also possible to claim extensions of range associated with the abundance of man-made habitats in south-east England (Carr 1986), the area most likely to be colonised by migrants from the Continent. However, in the same way that we lack survey data for ancient woodland ponds, we have no national overview of the faunas of reservoirs, of canals, or of ponds associated with the extractive industries. The 'piecemeal' approach of the recording scheme must be backed up by directed surveys.

Table 1. Post-glacial changes and their effects on water beetles

Events	Effects
Weichselian glaciation	Loss of entire fauna
Palaeolithic rapid warming	Colonisation
Mesolithic	
English Channel & separation of Ireland	Restricted British fauna More restricted Irish fauna
Maximal temperature followed by decline	Relict species Apparent southern distributions with northern refugia
Neolithic	
loss of wildwood	Loss of woodland species
Roman occupation to 1850s	
Fen and mere drainage Canal construction	Conversion of stagnant habitats to running water
Middle Ages	
Turbaries	Species restricted to Broadland
Little Ice Age	Extinctions and most extreme restrictions on refugia
Enclosure Acts	
Planned countryside	Apparent northern distributions with southern refugia or disjunct distributions
Industrialisation	
Water power	Damming of rivers
Fuel power	Acidification
Pollution	Loss of river species
Urbanisation	Lowland habitat loss
Canalisation	Loss of river species
Reservoirs	Stagnant, fluctuating open water habitats suitable for opportunistic species
Extractive industries	Continuum of base-rich open water habitats suitable for opportunistic species
Agricultural intensification	Grazing fen ditches changed to arable fen ditches Reduction in number of ditches Loss of peat substratum Eutrophication Marine incursions
War	
Training	Survival of habitats
Coastal defence	New coastal lagoons
Airfields	Habitat loss
Oil 'infrastructure'	Loss of coastal habitats
Afforestation	Loss of deep peat Loch acidification Stream siltation Increase in shaded habitats
Affluence	Loss of rich fens for horticultural peats Acidification from car emissions 'Manicuring' of habitats

Changes in communities

Water beetle recording lends itself to site-specific lists. There has been a shift in emphasis from species mapping to analysis of site lists, their ordination, classification and ranking in terms of conservation quality (eg Eyre, Ball & Foster 1986; Foster *et al.* 1990). Species lists within a region can be objectively divided into community types by use of

TWINSPAN (Hill 1979a). The factors dictating the differences between groupings can be identified independently of site descriptions by what is known of the ecological preferences of the water beetles; the program DECORANA (Hill 1979b) can be used to identify a hierarchy of ecological factors. The goodness-of-fit of a site list to one of a series of community types (its 'typicalness') can be tested using the standard deviations of the DECORANA ordination score to measure the site list's distance from the centroid of each group (Eyre & Rushton 1989). The priority for recording has thus changed from individual records of rare species to the need for complete lists that can be subjected to multivariate analysis.

Foster *et al.* (1990) recognised eight community types in part of the drainage system running into The Wash. They used 'species-quality score', an index of the average rarity status of all species found at a site, to rank site lists within listed groups according to community. On this basis it was recognised that the highest-quality, small farm drains were those that were neglected, whereas the best of the larger ditches, such as those for which internal drainage boards are responsible, are those that are regularly cleared of vegetation; the converse was also true – well-managed small ditches and neglected larger drains had low species quality. The fate of one site, the Cross Drain, Lincolnshire, could be traced from its 'discovery' as a high-quality site, through

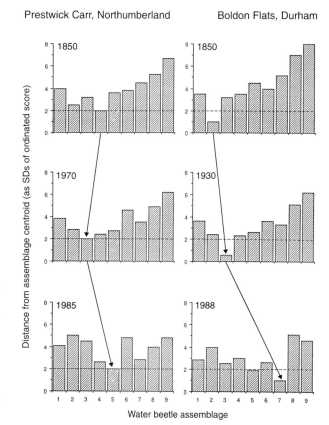

Figure 2. Changes in the water beetle assemblages at two sites in north-east England, as indicated by the best fit (in terms of standard deviations of ordination score from DECORANA) to the nine habitat groups recognised

vegetation clearing operations, deepening to receive a new water supply for irrigation purposes, and final receipt of that water from the River Welland; its species quality declined at each step and at the last it could be reclassified into a different community type.

It is also feasible to use classification and ordination to detect changes over a longer period. Eyre *et al.* (1986) recognised nine community types in north-eastern England. Reference to species lists recorded in the notebooks of T J Bold in 1850 and J Omer-Cooper in 1930 allowed M D Eyre (unpublished) to trace changes in community type and hence the most likely habitat type at several sites (Figure 2). Thus, in 1850 Boldon Flats in County Durham was probably a large bare pond; in 1930, according to the list of water beetles alone, it was a large muddy pond; and in 1988 it was a temporary pool with water beetles characteristic of such habitats. In 1850 Prestwick Carr in Northumberland was probably a *Sphagnum* mire, in 1970 it was a muddy pond, and in 1985 the same area was a lowland marsh.

This site approach is of limited value, but it is possible to develop a collective estimate for national or regional trends in invertebrate assemblages. These trends could be related to climate change or be associated with acidification or nitrogen enrichment. Recovery from pollution events is easily established on the basis of species quality.

REFERENCES

Balfour-Browne, W.A.F. 1940; 1950; 1958. *British water beetles.* 3 vols. London: Ray Society.

Bameul, F., Foster, G.N. & Holmen, M. 1982. Données récentes sur la géonemie et l'écologie de *Agabus (Gaurodytes) melanarius* (Col. Dytiscidae) en France, au Danemark, et en Grande-Bretagne. *L'Entomologiste,* **38**, 159–172.

Bilton, D.T. 1988. A survey of aquatic Coleoptera in central Ireland and the Burren. *Bulletin of the Irish Biogeographical Society,* **11**, 77–94.

Carr, R. 1986. The effects of human activity on the distribution of aquatic Coleoptera in south-eastern England. *Entomologica Basiliensia,* **11**, 313–325.

Eyre, M.D. & Rushton, S.P. 1989. Quantification of conservation criteria using invertebrates. *Journal of Applied Ecology,* **26**, 159–171.

Eyre, M.D., Ball, S.G. & Foster, G.N. 1986. An initial classification of the habitats of aquatic Coleoptera in north-eastern England. *Journal of Applied Ecology,* **23**, 841–852.

Foster, G.N., Foster, A.P., Bilton, D.T. & Eyre, M.D. 1990. Classification of water beetle assemblages in arable fenland and ranking of sites in relation to conservation value. *Freshwater Biology,* **22**, 343–354.

Hill, M.O. 1979a. *TWINSPAN – a FORTRAN program, for arranging multivariate data in an ordered two-way table by classification of the individuals and attributes.* Ithaca, New York: Section of Ecology & Systematics, Cornell University.

Hill, M.O. 1979b. *DECORANA – a FORTRAN program for detrended correspondence analysis and reciprocal averaging.* Ithaca, New York: Section of Ecology & Systematics, Cornell University.

Rackham, O. 1986. *The history of the countryside.* London: Dent.

Evidence for the effects of atmospheric pollution on bryophytes from national and local recording

K J Adams[1] and C D Preston[2]

[1] Polytechnic of East London, Romford Road, London E15 4LZ
[2] Biological Records Centre, Environmental Information Centre, Institute of Terrestrial Ecology, Monks Wood Experimental Station, Abbots Ripton, Huntingdon, Cambs PE17 2LS

INTRODUCTION

The effects of atmospheric pollution on the lichen flora of the British Isles are well known. They have not only generated a large body of scientific literature, but are familiar to the general public. By contrast, less attention has been paid to the effects of atmospheric pollution on bryophytes. A study of the relevant bibliographies suggests that in the 1980s papers on the effects of pollution on lichens outnumbered those on bryophytes by about four to one. Although these bibliographies are not strictly comparable, and probably exaggerate the preponderance of work on lichens[1], this figure is an indication of the relative attention paid to the two groups. Furthermore, the effects of atmospheric pollution on bryophytes are scarcely appreciated outside scientific circles.

The reason for the marked disparity between the work of bryologists and lichenologists probably lies in the larger number of lichen species available for use in pollution studies. Bryophytes can be shown to demonstrate the same range in apparent susceptibility to atmospheric pollution as can lichens. They are, perhaps, rather more conspicuous and slightly easier to identify; they would therefore be more suitable as monitors of pollution, other things being equal. However, the much larger suite of lichens which can be used in pollution studies is an advantage which greatly outweighs any difference there might be in ease of identification. The numerical difference in epiphytes, an ecological group which tends to be particularly susceptible to atmospheric pollution, is illustrated in Table 1. This difference probably reflects the greater tolerance of lichens to xeric conditions.

In this paper, we attempt to review the evidence for the effect of atmospheric pollution on bryophytes in Great Britain which can be obtained from national distribution studies (Preston), and to compare it with evidence from intensive recording at the local scale in a highly polluted area, London and Essex (Adams).

THE POLLUTANTS

The deleterious effects of air pollution on cryptogams was first noticed in the mid-19th century. It is not surprising that the earliest observations were made in 'grim, flat, smoky Manchester' (Grindon 1859a[2]), the town which was such a potent symbol of industrial growth to the early Victorians (eg Disraeli 1844; cf Briggs 1968). Writing in *The Manchester Flora*, Grindon (1859b) commented that 'the majority of those [lichens] enumerated are not obtainable nearer than on the high hills beyond Disley, Ramsbottom, Stalybridge, and Rochdale, and even there the quantity has been much lessened of late years, through the cutting down of old woods, and the influx of factory smoke, which appears to be singularly prejudicial to these lovers of pure atmosphere'.

Table 1. A comparison of the number of epiphytic mosses, liverworts, total bryophytes and lichens in two regions and on one host genus

All species recorded on living bark are included, whether or not they occur in other habitats. The number of epiphytes is also expressed as a percentage of the corresponding species total for the two regions. The numbers are derived from Bowen (1968, 1980), Hill (1988), Jones (1952b, 1953), Pentecost (1987) and Rose (1974).

The areas from which bryophytes and lichens are compared in north Wales are not absolutely identical, as data for bryophytes have been extracted for vice-counties 48, 49 and 52 whereas for lichens the area covered is modern Gwynedd, which covers vice-county 49, 52, much of vice-county 48 and a small area of vice-county 50.

	Mosses		Liverworts		All bryophytes		Lichens	
	Total	Epiphytes	Total	Epiphytes	Total	Epiphytes	Total	Epiphytes
Berkshire and Oxfordshire	305	60 (20%)	86	12 (14%)	391	72 (18%)	366	189 (52%)
North Wales	497	57 (11%)	208	32 (15%)	705	89 (13%)	851	317 (37%)
On oak (*Quercus* spp.)		48		17		65		303

Although evidence of the effects of pollution gradually accumulated, the lack of recording gauges for sulphur dioxide (SO_2) delayed critical study of the causal agents until 1958 (James 1973). Thus, Jones (1952a,b) was still discussing the effects of pollution in terms of 'smoke'. Since 1958, fieldwork, transplants and experimental studies on cryptogams have demonstrated that SO_2 is the most important of the atmospheric pollutants. Rydzak's (1959) alternative hypothesis, that the disappearance of cryptogams from the vicinity of towns was caused by adverse micro-climatic conditions, particularly lower humidity, was not supported by Skye (1958) or LeBlanc (1961), who both found a correlation between SO_2 levels and epiphyte abundance in areas of high humidity. Rydzak's hypothesis has subsequently been rejected by both bryologists and lichenologists (LeBlanc & Rao 1973a; Coppins 1973).

In both Europe and North America, the major source of SO_2 is the combustion of the fossil fuels, coal and oil, for domestic heating, electricity generating, and industrial uses such as oil refining (Saunders & Wood 1973). Natural gas, by contrast, contains little sulphur. Industrial processes, such as the manufacture of bricks and the sintering of iron ore, can be significant point sources of SO_2 (Warren Spring Laboratory 1972; Rao & LeBlanc 1967; Department of the Environment 1980), although their overall contribution is much less than that of fossil fuels. The amount of SO_2 emitted by natural phenomena, such as volcanic activity or forest fires, is negligible.

The estimated annual mean concentration of SO_2 in the atmosphere in 1987 is shown in Figure 1. The important feature of the map is the distribution of highly and less highly polluted areas; absolute values for a single year are less significant as they have decreased markedly in recent years (Munday 1990).

SO_2 accumulates on surfaces as the dry gas, dissolves in water as the highly toxic sulphite and bisulphite ions, or becomes oxidised to SO_3, dissolving in fog and rain drops as sulphuric acid. Winner, Atkinson and Nash (1988) estimated that on a dry weight basis the leaves of mosses absorb SO_2 at least 100 times more effectively than those of vascular plants. In addition to its direct effects, SO_2 modifies the habitat of epiphytes by acidifying bark and reducing its buffering capacity (Coker 1967).

There are numerous other atmospheric pollutants, which are less well documented than SO_2 and are not considered in detail here. However, it is worth pointing out that the concentrations of many (including nitrogen oxides, ozone and chlorofluorocarbons) have continued to increase in recent years, whereas concentrations of SO_2 have fallen in many areas. Any effects which these other pollutants have on bryophytes may become increasingly apparent.

The effect of heavy metal pollution on bryophytes lies outside the scope of this paper. There has been a number of studies on bryophytes in the vicinity of point sources of pollution. The extensive use of bryophytes for monitoring levels of metal pollutants is reviewed by Burton (1986).

THE NATURE OF EVIDENCE FROM FIELD STUDIES

In seeking field evidence about the possible effects of atmospheric pollution on bryophyte distribution, we are attempting to answer four questions.

1. Is the current distribution of a species correlated, either positively or negatively, with that of a pollutant?

2. Does any correlation in space extend to a correlation in time? There has been temporal as well as spatial variation in the concentration of SO_2. If SO_2 is an important factor determining the distribution of a species, we would expect changes in its distribution in response to these variations in concentration. There might, however, be a lag period before the response. The rate at which a species can colonise newly available habitats will depend on the availability of source populations and the mobility of the species. The effects of high concentrations of a pollutant might remain even after those concentrations have fallen. For example, if high levels of SO_2 have resulted in bark acidification, old acidified bark may persist for a period after the deposition levels have dropped.

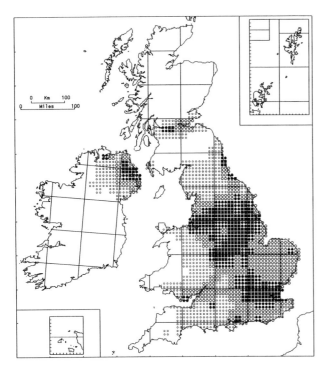

$■$ >32 μg SO_2 m^{-3}
$●$ 24–32 μg SO_2 m^{-3}
$○$ 16–24 μg SO_2 m^{-3}
$○$ 8–16 μg SO_2 m^{-3}

Figure 1. The estimated mean annual concentration of sulphur dioxide in the atmosphere in 1987, expressed as the mass of SO_2 m^{-3}, in the United Kingdom. Emissions of sulphur dioxide have fallen sharply during the 30 years of the BBS mapping scheme. Map based on information from Warren Spring Laboratory.

3. Do correlations observed in Britain also apply in other countries? There is detailed evidence of the past and present distribution of bryophytes in other countries, particularly in north-west Europe. This evidence has been summarised in map form for liverworts in Belgium (Schumacker 1985) and mosses in the Netherlands (Touw & Rubers 1989). There are also studies of bryophytes around point sources of SO_2 pollution in the USA and Canada (reviewed by Winner 1988).

4. Is there another explanation for the past and present distribution of species? Many land use changes have taken place in the areas of greatest SO_2 concentration. Obvious examples include the destruction of semi-natural vegetation by urban development, changes resulting both from agricultural intensification (eg land drainage, destruction of hedgerows, use of artificial fertilizers), and from the increasing predominance of arable agriculture at the expense of pasture. Many species have declined as a result of these changes, one of the more striking examples being *Splachnum ampullaceum*, a moss which grows on mammal dung in fens, moorlands and bogs, and is now extinct in many south-eastern counties.

Answers to these four questions are only likely to be obtainable for taxa which are (or were) common or widespread, occurring in areas with differing concentrations of pollutants. The susceptibility of rare species which have always been confined to a few localities is unlikely to be detected by these methods, although one might suspect it if the trends in the distribution of a rarer species are consistent with those shown by related, commoner species for which more data are available.

Field recording can establish correlations between species distributions and levels of pollutants, and these correlations suggest hypotheses which can be tested by experimental methods. It is always important to bear in mind the fact that our hypotheses based on observed correlations can only be proved by rigorous experimental testing. However, it is also important that those involved in the acquisition and analysis of records should not underestimate the importance of the generation of hypotheses in the scientific process, or allow it to be underestimated by others.

EVIDENCE FROM THE QUATERNARY SUB-FOSSIL RECORD

The most devastating effect of atmospheric pollution on the semi-natural vegetation of the British Isles has been revealed primarily by evidence from the sub-fossil record (Press, Ferguson & Lee 1983). Examination of the peat deposits below the southern Pennine blanket bogs shows that *Sphagnum* species were formerly a much larger component of the blanket bog vegetation than they are today. The two species which have in the past been major peat formers, *S. imbricatum* and *S. magellanicum*, are now completely absent from the area, and the bogs are currently dominated by cotton-grass (*Eriophorum vaginatum*). The appearance of soot deposits in the peat profile at the time of the Industrial Revolution is followed by the almost complete disappearance of *Sphagnum* remains. Suggestions that the reduction in *Sphagnum* cover was caused by atmospheric pollution are strongly supported by experimental studies, which demonstrate that artificial acid rain reduces the growth and photosynthesis of *Sphagnum*, and proves lethal to the more sensitive species, at concentrations which do not affect vascular plants. The most resistant of the species tested experimentally, the flush-dwelling *S. recurvum*, is the most widespread species in the southern Pennines today (Ferguson, Lee & Bell 1978; Ferguson & Lee 1980).

Sulphur dioxide concentrations are much lower in the southern Pennines today than they were in the 19th century. Nevertheless, *Sphagnum* continues to be scarce even in areas where suitable habitats are apparently available for colonisation, and attempts to transplant *Sphagnum* to these sites only succeeded in establishing the most pollution-tolerant of the species investigated, *S. recurvum* (Ferguson & Lee 1983). It seems likely that deposition of atmospheric nitrogen, which has increased four-fold in the Manchester area since the 1860s, now inhibits recolonisation by *Sphagnum* species (Ferguson *et al.* 1984; Lee 1986; Press, Woodin & Lee 1986).

The sub-fossil record of *Sphagnum* is outstandingly rich; no other bryophyte genus is anything like as well represented (Dickson 1973). Nevertheless, sub-fossils of some other moss genera provide valuable distributional evidence for the period before botanical recording and before widespread atmospheric pollution (see, for example, the discussion of *Antitrichia curtipendula* below). Liverworts are, by contrast, very poorly represented in the sub-fossil record.

EVIDENCE FROM STUDIES OF NATIONAL DISTRIBUTION

The British Bryological Society's mapping scheme was launched in 1960, under the supervision of Dr A J E Smith. The scheme has concentrated on field survey, designed to establish the distribution of British bryophytes at the 10 km square scale, but significant historical records have also been extracted from literature sources and herbarium specimens. Over 750 000 records collected by the scheme were added to the computer database at the Biological Records Centre (BRC) between 1985 and 1989, and these have now been summarised as distribution maps. The first volume of a three-volume *Atlas of bryophytes of Britain and Ireland*, covering liverworts, has been published (Hill, Preston & Smith 1991); the moss volumes should be published by 1993. The draft maps prepared for the *Atlas* have been used in this preliminary study. Once work on the *Atlas* is completed, a more detailed analysis of the dataset will be possible.

Decreasing species

Species which have been most adversely affected by atmospheric pollution at the national scale (judged by the four criteria outlined above) are listed in Table 2. There is evidence that all these species formerly grew in areas which are now characterised by high, or moderately high, mean annual concentrations of SO_2, but are now absent from them or are much rarer in those areas than they were. The 20 taxa listed comprise approximately 2% of the British bryophyte flora.

Table 2. Species which have apparently been most adversely affected by atmospheric pollution at the national scale

Epiphytes (ie plants growing on the bark of trees and shrubs)	**Epiliths** (ie plants growing on rocks and boulders)
* *Cryphaea heteromalla*	*Grimmia affinis*
† *Frullania dilatata* (also grows on rocks and boulders in the west)	*G. decipiens*
	G. laevigata
* *Neckera pumila*	*G. orbicularis*
* *Orthotrichum lyellii*	*G. ovalis*
* *O. obtusifolium*	**Species which can grow as epiphytes or epiliths**
* *O. schimperi*	
* *O. speciosum*	**Antitrichia curtipendula*
* *O. stramineum*	**Leucodon sciuroides*
* *O. striatum*	
* *O. tenellum*	
* *Tortula laevipila*	
* *Ulota crispa* var. *crispa*	
* *U. crispa* var. *norvegica*	

† Has declined in Belgium (Schumacker 1985)

* Has declined in the Netherlands (Touw & Rubers 1989). Evidence from Dutch sources is not available for the *Grimmia* species; *G. laevigata* is the only species recorded in the Netherlands and it is restricted to two recently discovered localities

Most of the species listed characteristically grow as epiphytes, only occasionally occurring on other substrates. These are the 'obligate epiphytes' in the sense that Smith (1982) uses the term, but would more accurately be described as predominantly epiphytic. The sensitivity of some epiphytic bryophytes to pollution is well known (cf Rose & Wallace 1974). Epiphytes which have decreased in areas of high SO_2 concentrations (ie areas shown on Figure 1 with over 24 µg SO_2 m^{-3} in 1987) include *Ulota crispa* (Figure 2); species which have decreased even in areas of moderately high SO_2 concentrations (ie 8–24 µg SO_2 m^{-3} in 1987) include *Orthotrichum striatum* (Figure 3). Both species have shown a similar decline in the Netherlands, and *Ulota crispa* has

● Post-1950
○ Pre-1950

Figure 2. Distribution of *Ulota crispa* in Britain and Ireland. Records of both var. *crispa* and var. *norvegica* are included on the map. Many of the scattered post-1950 records in eastern England represent populations discovered in the last ten years

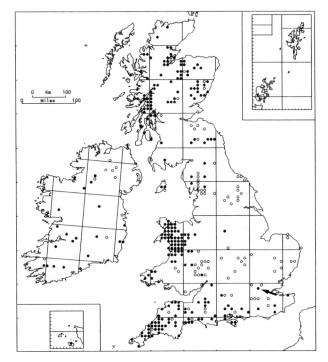

● Post-1950
○ Pre-1950

Figure 3. Distribution of *Orthotrichum striatum* in Britain and Ireland. The recently discovered Hampstead population is indicated by an arrow

also been identified as a particularly sensitive species in Canada (Rao & LeBlanc 1967). *Orthotrichum striatum, Ulota crispa* and most of the other epiphytes listed in Table 2 still survive in areas of low SO$_2$ concentrations in the north and west. However, there is a very small group of species with a more continental distribution which have never been recorded in the west. These plants of the central and eastern counties have been particularly severely affected, as they have been virtually eliminated from areas of moderately high pollution in England and survive only in eastern Scotland. *Orthotrichum obtusifolium* (Figure 4) is the best example; it was extinct in England from 1921 until 1989, when a single tuft was discovered on a roadside elder (*Sambucus nigra*) in Norfolk. *O. obtusifolium* is a frequent epiphyte in Canada, where studies have shown that it decreases greatly in frequency and fertility towards the polluted centre of Montreal (LeBlanc & Rao 1974; LeBlanc & De Sloover 1970) and is damaged by both SO$_2$ and fluoride pollution when transplanted from unpolluted to polluted sites (LeBlanc, Comeau & Rao 1971; LeBlanc & Rao 1973b). The rarer *Orthotrichum speciosum*, extinct in Sussex and Yorkshire, but still present in Scotland, probably belongs in the same category.

The epilithic *Grimmia* species listed in Table 2 are species of well-illuminated, dry, acidic or basic rocks, often at low altitude. Their decline has not usually been attributed to atmospheric pollution, but Hill (1988) points out that at least some epiliths are likely to be affected by pollution, and it is difficult to suggest another explanation for the decline. Exper-

imental studies of the susceptibility of these species to atmospheric pollutants are clearly desirable.

Leucodon sciuroides can grow either epiphytically or epilithically. In some counties of south-east England (eg Cambridgeshire, Surrey), it is no longer present as an epiphyte, but survives on church walls and churchyard monuments, usually, if not exclusively, on calcareous stone. This restriction might be the result of atmospheric pollution alone, but it is also possible that churchyards provide a sanctuary from other potentially adverse changes, such as the eutrophication of substrates by atmospheric drift of agricultural fertilizers.

The remarkable decline of *Antitrichia curtipendula* in England has often been commented on, and has been attributed (by Rose & Wallace 1974) to the particular sensitivity of this species to SO$_2$, coupled with the felling of ancient trees. *Antitrichia* is an easily recognised species, well represented in the fossil record. It was abundant in the west in the Late Devensian, and spread rapidly to colonise eastern England in the early Flandrian (Dickson 1973). It is found in Bronze Age, Roman and Saxon archaeological sites in Hampshire, Oxfordshire, Suffolk, Huntingdonshire, Lincolnshire, Nottinghamshire and south-east Yorkshire (Dickson 1973, 1981; Stevenson 1986). Its decline in England may have begun before the onset of widespread SO$_2$ pollution; the decline has been so extensive that it is doubtful whether it can have been caused by atmospheric pollution alone, although this is almost certainly one of the factors responsible.

Increasing species

It is much more difficult to establish that a species is increasing than it is to demonstrate a decrease. This difficulty is especially true for a relatively inconspicuous group such as the bryophytes, where one can rarely be certain that the absence of earlier records indicates the absence of the plant rather than the fact that it was overlooked. However, there is evidence to suggest that three members of the Dicranaceae, which grow epiphytically on acid bark, have increased: *Dicranoweisia cirrata* in the late 19th and early 20th centuries and *Dicranum montanum* and *D. tauricum* (synonym: *D. strictum*) more recently (Rose & Wallace 1974; Smith 1978). *Ptilidium pulcherrimum* is an epiphytic liverwort which may have increased (Wallace 1963), although the evidence is not conclusive; it is interesting to note that Rao and LeBlanc (1967) found that this bryophyte was the most tolerant to SO$_2$ in the vicinity of a Canadian iron sintering plant.

EVIDENCE FROM LOCAL STUDIES IN LONDON AND ESSEX 1800–1990

The rise and fall of atmospheric pollution in London

With the widespread use of coal for domestic heating in the mid-19th century, and the development of

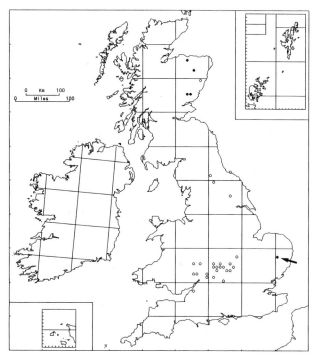

• Post-1950
○ Pre-1950

Figure 4. Distribution of *Orthotrichum obtusifolium* in Britain and Ireland. The recent Norfolk record is indicated by an arrow

coal-burning power stations, in response to massive population growth, London became a potent source of atmospheric pollution, dominated by suspensions of smoke and soot particles, and in particular SO_2. The virtually pristine air of the early 1800s began to deteriorate rapidly soon after that date, and by the 1870s many species of bryophytes and lichens were already becoming scarce in London. The area to the north-east of the metropolis was also affected, as pollutants were dispersed by the predominantly south-westerly winds. As Crombie (1885) noted, 'the paucity of lichen-growth [in Epping Forest] . . . is attributable to the atmosphere being in some states of the weather more or less impregnated with smoke from the increased number of human habitations on the outskirts of the Forest, the acids contained in which are most destructive to lichens. Add to this that the direction of the prevailing winds being from the SW, the smoke and fogs of London and its sub-urbs extend their deleterious influence at certain seasons of the year to the nearer portions of the Forest, and even considerably beyond. . .'.

Despite the toxicity of SO_2 to vegetation and the dis-astrous effects of sulphuric acid on stonework and human lung tissue, the initial concerns were with the control of smoke emissions. Enormous quantities of particulate matter were being belched into the atmosphere of London by domestic coal-burning fires. The annual-averaged smoke concentration at ground level in London for 1954, for example, was some 250 µg m^{-3} (Warren Spring Laboratory 1972), and the smoke plume drifted downwind far out into Essex, coating vegetation and buildings with a layer of sticky black grime. The Clean Air Acts of 1956 and 1968 only legislated against smoke emissions. They were very effective, however, and rapidly cured the smoke problem, the last severe smog in London being recorded over the winter of 1962–63 (Chandler 1976). SO_2 levels in central London during the period 1954–64 were typically around 570 µg m^{-3} (0.2 ppm) during the winter months, peaking as high as 5700 µg m^{-3} (2.0 ppm); the corresponding figures for sulphuric acid were 10 µg m^{-3} and 700 µg m^{-3} (Lawther 1965).

Fortuitously, the change-over to less smoky, more efficient, domestic combustion systems coincided with the transition from coal to solid fuels with a lower sulphur content, and a gradual decline in atmospheric SO_2 concentration in London and East Anglia has taken place since the late 1950s. Although, in contrast, industrial/power station output of SO_2 in London continued to rise, the increase was offset by the installation of tall chimneys that disperse the SO_2 and sulphuric acid in the prevailing south-westerly winds, high enough up into the atmosphere for most of it to avoid London and the eastern counties, exporting it instead to Scandinavia as acid precipitation.

The greatest levels of atmospheric SO_2 near ground level do not occur during the coldest months when the output is greatest, but during the autumn when

inversions entrap the SO_2 from domestic sources, and when bryophytes and lichens are in their most active phase of growth. The hot rising plumes from tall power station chimneys 'punch' holes through the inversion layers, allowing long-range dispersal despite the presence of static air at ground level.

The increasing use of sulphur-rich oil for domestic heating since 1950 is thought to have been at least partly responsible for a temporary rise in SO_2 levels in London between 1961 and 1965. Subsequent legislation to reduce the sulphur content of heating oils, and the increasing use of natural gas have pro-moted a gradual decline in domestic SO_2 output. Since 1965 the rate of fall in the annual-averaged atmospheric SO_2 concentration in London has continued more or less linearly at approximately 60 µg m^{-3} per decade (Figure 5). This gradual fall has probably also been assisted by the closing down of several of the older, less efficient, coal-burning power stations in London, and a general decline in heavy manufacturing industry.

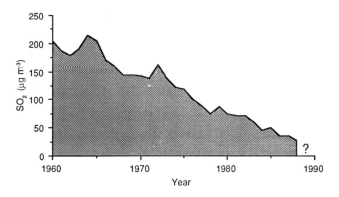

Figure 5. Mean trends in the March–April annual-averaged atmospheric SO_2 concentrations (µg m^{-3}) for a selection of sites in London from 1959–60 to 1987–88. The rate of fall in concentration approximates to 60 µg m^{-3} per decade

Despite the gradual decline in SO_2 output in London since the late 1950s, levels remained toxic to many lichens and bryophytes in London and south-west Essex until well into the 1970s. Thus, in 1971–72, SO_2 levels reached a mean winter value of 170–180 µg m^{-3} in the south-west corner of vice-county 18 (South Essex), falling to c120 µg m^{-3} in an arc from Chingford to Thurrock; to about 70 µg m^{-3} in an arc to the north-east of London passing through Chelmsford; and to around 60 µg m^{-3} at the Essex/Suffolk border – only the extreme coastal areas of Norfolk receiving less than 40 µg m^{-3}. It was not until the mid-1970s that there was any noticeable effect on SO_2-sensitive lichens, and probably not until the early to mid-1980s on sensitive bryophytes, as a result of the gradual fall in SO_2 levels in London and Essex, and the bryophyte flora was still in obvious decline during the period 1967–74 when a baseline survey was carried out for the *Flora of Essex* (Jermyn 1974).

Even as late as 1984–85, however, winter levels in central London were around 130 µg m^{-3} whenever prevailing easterly winds brought in SO_2 from power stations to the east of London, and short-term episodes peaked in excess of 1000 µg m^{-3} (Harrop, Laxen & Daunton 1985). These seasonal and short-term fluctuations make it extremely difficult to correlate SO_2 levels with damage to bryophytes, as single short-term pulses could well be lethal to some species against a background level which they might otherwise tolerate.

The decline of the bryoflora

The early history of the decline of bryophyte (and lichen) species in the area north-east of London is fragmentary and largely dependent on three collectors: Edward Forster, who collected mainly in the Epping Forest and Walthamstow areas around 1800; Ezekiel G Varenne, mainly in the Kelvedon area from 1860 to 1876; and Frederick Y Brocas in the Saffron Walden area around 1847. Forster's herbarium, in particular, is crucial to our understanding of the lethal effects of air pollution on bryophytes. Around 1800 he collected extensively in the Epping Forest area, finding many species that have not otherwise been recorded in Essex; he effectively set a baseline for future observations.

The decline of epiphytic bryophytes, and particularly lichens (Rose & Hawksworth 1981), has been extensively highlighted nationally. In Essex about 20 bryophyte species have been severely affected, notably members of the Orthotrichaceae. *Neckera pumila, Orthotrichum schimperi, O. sprucei, O. striatum, Ulota crispa* var. *crispa* and *Zygodon conoideus* are all extinct in Essex, and *Cryphaea heteromalla, Leucodon sciuroides, Orthotrichum lyellii, Tortula papillosa* and *Ulota crispa* var. *norvegica* are confined to isolated localities in north Essex, though they were formerly widespread. Other species have been severely restricted by pollution in south-west Essex, but are reasonably abundant further out. *Frullania dilatata, Radula complanata, Porella platyphylla* and *Isothecium myurum* are all now absent in south-west Essex, and *Neckera complanata* only survives as isolated scraps deep in sheltered woodland. These species appear to have become obligate calcicoles, following fairly closely the distribution of chalk and chalky boulder clay deposits in north-west Essex. This pattern may be an artifact, more related to SO_2 levels than substrate acidity, however. Curiously, *Lejeunea cavifolia* persists on a single tree by a deep ditch in Running Water Woods, Upminster, although it now appears to be extinct elsewhere in Essex.

Of the epiphytes that have survived the onslaught of SO_2 in south-west Essex, *Lophocolea heterophylla, Brachythecium rutabulum, Rhynchostegium confertum, Eurhynchium praelongum* and *Isothecium myosuroides* provide the main cover on tree bases and stumps, with *Tetraphis pellucida* and *Orthodontium lineare* taking over in acid areas. *Hypnum mammillatum* is surprisingly abundant in Monks Wood, Epping Forest, seeming to be no more sensitive to pollution than corticolous *H. cupressiforme* var. *cupressiforme* though more so than *H. cupressiforme* var. *resupinatum*.

Although *Dicranoweisia cirrata* has been compared with *Lecanora conizaeoides* in its ability to colonise the so-called lichen desert, it is far more sensitive to pollution than *Lecanora* in eastern England. *D. cirrata* survived in the urban outskirts of London during the worst of the pollution, though was seldom found fruiting, but in the most heavily polluted areas virtually all the epiphytic species were wiped out. *Hypnum cupressiforme* var. *resupinatum* and *Ceratodon purpureus* were probably the most resistant epiphytic mosses (on tree bases) in heavily polluted areas. *Dicranoweisia* may have been able to establish itself by leaf gemmae, which it produces copiously in eastern England, in areas too polluted for the protonema to establish from spores.

A few species only able to persist intermittently between intense episodes of pollution in eastern England may have been continuously replenished by spores or leaf fragments blown in on the wind, or carried by birds, from the Continent. The sporadic and ephemeral occurrence of *Ptilidium pulcherrimum, Dicranum flagellare, D. tauricum* and *Ulota crispa* var. *norvegica* in many localities suggests some of them may have arrived in this way.

By looking at the sequence of extinctions against rising levels of SO_2, and the present-day distribution of species in conjunction with known SO_2 levels between 1950 and 1970, it is possible to draw up a table of bryophyte sensitivities equivalent to that of the Hawksworth and Rose (1970) scale for lichens. Table 3 is based on the deduced extinction sequence for Epping Forest and the surrounding counties. This scale must be interpreted with caution, however, as the position of a species on the scale will depend on habitat humidity, and, as in the case of the lichen scale, the acidity of the bark substrate. Thus, species of *Ulota, Orthotrichum, Zygodon viridissimus, Frullania dilatata, Tortula laevipila* and *Metzgeria furcata* appear first on neutral barks, such as those of field maple (*Acer campestre*), crack willow (*Salix fragilis*) and elder.

In addition to the familiar pollution-sensitive epiphytic bryophytes, numerous terrestrial and saxicolous species have also been exterminated or rendered infertile by atmospheric pollution in many parts of Britain. Gilbert (1971) noted that in the worst affected areas of Newcastle-upon-Tyne (>170 µg m^{-3} SO_2) only *Ceratodon purpureus* and *Bryum argenteum* appeared able to survive in abundance. Further out, as levels fell to between 130 and 70 µg m^{-3} SO_2, *Tortula muralis* and *Bryum capillare* were added to the list; and *Grimmia pulvinata* and *Orthotrichum diaphanum* appeared on artificial substrates as the level fell to between 50 and 40 µg m^{-3}. This describes graphically the situation that prevailed in London and the inner suburbs until the early 1980s.

Even outside the London suburbs, where the bryophyte flora retained much of its diversity, there

Table 3. Mosses and liverworts found as epiphytes on the trunks, bases, crotches, stumps or roots of trees that showed poor or restricted growth during the maximum phase of SO_2 pollution, or were exterminated, with their approximate equivalent SO_2 thresholds. Based on historical records and field survey from 1967 to 1990 in Epping Forest and outer Essex

NB Some species may fare better on limestone than on trees, or may move up and down the scale depending on whether the bark is acid or neutral. Those species able to survive desiccation must also be expected to move up the table in areas of the country where the average humidity is higher than in the London/Essex area

Mean winter SO_2 (µg m^{-3})	Approximate pollution zone*		Last or only date recorded	
			Epping Forest	Outer Essex
'Pure'	9–10	#*Antitrichia curtipendula*	*c* 1800	1874
	9	*Orthotrichum sprucei*	–	1866
<30	9	*Orthotrichum schimperi*	–	1873
	9	*Orthotrichum tenellum*	–	1870
	9	*Ulota crispa* var. *crispa*	*c* 1800	1874
	9–8	*Orthotrichum striatum*	*c* 1800	1870
c 35	8	*Zygodon conoideus*	1885	1886
	8	*Neckera pumila*	1890	1874
	8–7	*Tortula papillosa*	–	1874
	7	*Ulota crispa* var. *norvegica*	–	:
c 40	7	#*Anomodon viticulosus*	1932	:
	7	#*Radula complanata*	*c* 1890	:
	7–6	*Leucodon sciuroides*	1885	:
	6	*Orthotrichum lyellii*	1898	:
	6	*Cryphaea heteromalla*	*c* 1800	:
c 50	6	*Frullania dilatata*	1923	:
	6	#*Homalia trichomanoides*	1973	:
	6	*Porella platyphylla*	*c* 1890	:
	6–5	*Isothecium myurum*	1885	:
	5	*Tortula laevipila*	*c* 1980	:
	5	*Neckera complanata*	:	:
c 60	5	*Zygodon viridissimus*	:	:
	5	*Orthotrichum affine*	:	:
	5	*Orthotrichum diaphanum*	:	:
	5–4	*Homalothecium sericeum*	:	:
	4	*Hypnum mammillatum*	:	:
	4	*Hypnum cupressiforme* var. *cupressiforme*	:	:
c 70	4	*Dicranum scoparium*	: extant	:
	4	*Isothecium myosuroides*	:	:
	4	*Bryum capillare*	:	:
	4–3	*Dicranoweisia cirrata*	:	:
	3	*Hypnum cupressiforme* var. *resupinatum*	:	:
c 125				
	3	*Lophocolea heterophylla*	:	:
*c*150	2–3	*Ceratodon purpureus*	:	:

Species which may be limited by factors other than specific sensitivity to SO_2
* Hawksworth and Rose (1970)

were specific casualties extending well out into East Anglia. Among terrestrial mosses, for example, *Rhytidiadelphus loreus*, last recorded in Epping Forest in 1912, and in Hatfield Forest in 1890, is now extremely rare in eastern England. Its decline was closely followed by that of *Rhytidiadelphus triquetrus*. Recorded as 'common' and collected fruiting in Epping Forest in 1800 (Forster), it had become 'very scarce' there by 1885 (J T English) and must have become extinct well before the 1960s when E Saunders noted its absence, as by 1967 it had also been exterminated over a wide area of Essex. It was last seen in Hatfield Forest at a single site in 1965, for example. Although *R. loreus* appears to be virtually extinct in eastern England today, *R. triquetrus* still occurs in numerous pockets, apparently as an obligate calcicole, in north and east Essex, Suffolk and Norfolk. In contrast *Rhytidiadelphus squarrosus* remained an abundant grassland species (eg in lawns), even in the outer London suburbs, during the maximum phase of pollution. These three species demonstrate just how different the sensitivity to SO_2 can be, even among species in the same genus.

Bartramia pomiformis and *Hylocomium splendens* are further examples of sensitive terrestrial mosses. They were last recorded in Epping Forest in 1931 and 1912, respectively. *Bartramia* is now only found in isolated pockets in north-east Essex, while *H. splendens* is only known in Essex from one site, near Colchester. Even *Tortula ruralis* is not found commonly on the ground until well out into Essex towards Colchester. So far these species have not yet shown any response to the post-1950 SO_2 decline (Table 4).

Table 4. Terrestrial mosses now absent in the south-west of Essex that are believed to have succumbed to SO_2 pollution, in approximately decreasing order of sensitivity

---

Rhytidiadelphus loreus

Hylocomium splendens

Bartramia pomiformis

Rhytidiadelphus triquetrus

Tortula ruralis

Hypnum cupressiforme var. *cupressiforme*

In contrast, *Rhytidiadelphus squarrosus*, *Pseudo-scleropodium purum*, *Calliergon cuspidatum*, *Hypnum jutlandicum*, *Brachythecium rutabulum*, *Brachythecium albicans*, *Amblystegium serpens* and *Eurhynchium praelongum* all seem to have maintained their distribution and abundance during the severest period of pollution, but, curiously, *Hypnum cupressiforme* var. *cupressiforme* is largely absent as a terrestrial moss far out into East Anglia.

Ctenidium molluscum, *Philonotis fontana*, the bog mosses *Sphagnum tenellum* and *S. subnitens*, and the liverworts *Scapania irrigua*, *S. nemorea* and *S. undulata* are also suspected to have been affected by acid rain in south-west Essex, but there may have been other factors contributing to the decline of these species.

Bryophyte species characteristic of brick and stone walls, tiled roofs and churchyard tombs have also suffered extensively in London and Essex from atmospheric pollution. *Rhynchostegium confertum*, *Amblystegium serpens*, *Tortula muralis*, *Ceratodon purpureus*, *Barbula convoluta*, *Bryum capillare*, *B. argenteum*, *B. caespiticium* and *Hypnum*

cupressiforme var. *resupinatum* seem to have withstood the worst phase of pollution in the outer London suburbs, but *Barbula revoluta*, *Barbula rigidula*, *Tortula intermedia*, *Orthotrichum anomalum*, *Barbula vinealis*, *Zygodon viridissimus*, *Schistidium apocarpum*, *Homalothecium sericeum*, *Grimmia pulvinata* and *Orthotrichum diaphanum* – roughly in that order of decreasing sensitivity – have been severely affected. All these species became virtually extinct in central London and the inner suburbs during the worst of the pollution. In the outer suburbs, *O. diaphanum* and *H. sericeum* survived as calcicoles on stone, the latter seldom fruiting, but were largely absent as epiphytes. Today *H. sericeum* is only found as tiny scraps on gnarled tree roots in Epping Forest, with the exception of a handful of leaning trees where it occurs in rain tracks on the trunks.

The wind of change

Although the prevailing south-westerly winds from London cast a pall of death on the bryophyte and lichen communities in eastern England for over 100 years, the decline in SO_2 levels now appears to be taking effect. Spores and other propagules carried in on the wind from south and west Britain, and from the Continent of Europe, are leading to a spectacular recovery and recolonisation of bryophyte (and lichen) species.

The lichens were the first to respond, with recolonisation of the inner suburbs of London by foliose species in the period 1970–80 (Rose & Hawksworth 1981), and further spectacular improvements since then (Hawksworth & McManus 1989). In Essex and Middlesex there has been a dramatic recolonisation by lichens on crack willow, sallow (*Salix* spp.) and oak (*Quercus* spp.). In the period 1987–89 the recolonisation got well under way in outer Essex, but in the autumn of 1989 and spring of 1990 foliose lichens (with *Parmelia sulcata* in the vanguard) began to appear in quantity in Epping Forest and Bedfords Park. Large thalli of *Parmelia perlata* were discovered on sallows (*Salix cinerea*) by Baldwins Pond in 1989 in Epping Forest, and on willow in Pond Wood, Middlesex, and Hatfield Forest, Essex. Even more spectacularly, they were discovered on crack willow just north of Hampstead Ponds, Hampstead Heath in central London, growing together with *Usnea inflata*, formerly regarded as a species of the western and southern coasts.

Bryophytes have been much slower to respond than lichens. Possibly either residual acid in tree bark held them back for several additional years or the airborne density of propagules was a limiting factor. Significant increases in the luxuriance of existing plants have been noted in Essex, possibly encouraged by the mild winters of 1988–89 and 1989–90, in addition to the decrease in rainfall acidity. *Isothecium myosuroides*, formerly confined to tree bases, now clothes the trunks, up to the crowns, of many pollards in Gurnon Bushes, north of Epping, and *Orthotrichum affine*, previously only ever found as tiny sterile scraps a few millimetres high, has suddenly become an abundant and luxuriant, characteristically fruiting plant, on elder in several areas. In Bedfords Park a colony of *Frullania dilatata* appeared on a young oak in 1989–90, the first time it has been seen in south-west Essex since 1926, and *Orthotrichum pulchellum* has been reported for the first time ever (1990) in north Essex.

Perhaps most spectacularly of all, however, has been the finding (Adams 1990) of several bryophyte species formerly regarded as extinct in London, all growing together on crack willows in the Hampstead Ponds and Highgate Ponds valleys, on either side of Parliament Hill on Hampstead Heath. Some ten tufts of fruiting *Orthotrichum striatum* were found in the spring of 1989 on the horizontal bough of a crack willow by Hampstead Ponds. Subsequently numerous fruiting tufts of *Orthotrichum affine*, three tufts of *Ulota crispa* var. *norvegica* – also in fruit and all on different trees – and two colonies of *Frullania dilatata* were all discovered on the same patch of willows, virtually within sight of the Royal Free Hospital. Over in the next valley was a large tuft of *Ulota phyllantha*, again on willow. These records suggest that these newcomers must have established themselves several years previously, possibly as early as 1985, and clearly indicate that not only has SO_2 pollution been brought firmly under control in London, but that other gaseous pollutants, in particular ozone and the nitrogen oxides (which are still known to be rising in concentration due to vehicle emissions), do not seem to have a toxic effect on either bryophyte growth or re-establishment.

Tortula intermedia is also making a comeback, having reappeared on both old and new tiled roofs in Loughton, Essex, in about 1986–87, rapidly forming large cushions. *Orthotrichum diaphanum* has become much more luxuriant and is appearing in fruit on concrete walls all over the London suburbs, and *Schistidium apocarpum*, together with *Orthotrichum anomalum*, is reappearing on limestone surfaces, as by Highgate Ponds on Hampstead Heath.

Although it would appear that SO_2 is toxic to many species of bryophyte, and in particular, in varying degree, to members of the Orthotrichaceae, it may not be so for all species of bryophytes. In addition to being sensitive to SO_2, some species may be affected by other pollutants. Indeed, some SO_2-resistant species may be succumbing to other pollutants. Two species which appear to exhibit an anomalous response to the decline in SO_2 levels are *Homalia trichomanoides* and *Anomodon viticulosus*. Both these species have continued to decline throughout eastern England, and only occur in isolated pockets on neutral bark substrates in micro-habitats with a high humidity. Most surviving colonies are still showing signs of tissue damage, and many of the shoots are dead. Could these two species be sensitive to nitrogen oxides, ozone, or perhaps ammonia, in addition to SO_2?

PHYSIOLOGICAL EFFECTS OF POLLUTANTS

Very little is known about the precise effects of SO_2 and sulphite on epiphytic bryophytes (and still less on terrestrial species). Most experimental work has been centred on the experimental gassing of established gametophytes or reciprocal transplant experiments in the field. Coker (1967) has shown, however, that SO_2 can acidify bark and reduce its buffering capacity. Rotten bark apparently has a higher buffering capacity, which may explain why this is preferentially colonised by epiphytic bryophytes in heavily polluted areas. Re-establishment could, however, be affected by poisoning of the protonemal stages (Gilbert 1968), or the antherozoids, or even the spores, or gametophyte stress could result in suppression of gametangia.

Experimental gassing of epiphytic bryophyte gametophytes with known concentrations of SO_2 has shown that they respond to SO_2 exposure by attempting to oxidise the sulphite to sulphate, the latter being some 30 times less toxic (Thomas 1961). Most sensitive species appear to use respiratory energy for this purpose, exhibiting a burst of oxygen uptake following exposure, but at the higher SO_2 levels are rapidly overcome, beginning to show signs of physiological damage (eg chlorophyll breakdown and a fall in respiration rate) within a few hours of exposure, followed rapidly by cell death (Syratt 1969). A particularly obvious reducing effect of sulphurous acid is the replacement of the Mg++ ion in the porphyrin ring of chlorophyll by 2H+, causing a bleaching of the characteristic green colour and formation of the grey phaeophytin.

Dicranoweisia cirrata has been shown by gassing experiments to be able to use photosynthetic energy to detoxify SO_2 and accumulate sulphate with relative impunity (Syratt 1969). In darkness, however, it has to rely on respiratory energy and appears to be no more efficient at accumulating sulphate than such sensitive species as *Ulota crispa*. It would appear, therefore, that an ability to sustain a high level of oxidative capacity to detoxify continuously absorbed sulphite may be the critical factor enabling a species to resist SO_2 poisoning, rather than any species-specific variation in sensitivity to accumulated sulphate. Thus, species capable of utilising light energy for this purpose are likely to gain a competitive advantage over those restricted by limited reserves of respiratory substrates. Furthermore, Syratt (1969) has shown that even resistant species suffer some chlorophyll breakdown on exposure to SO_2, and so a high chlorophyll content may be a further pre-requisite to resistance, enabling plants to survive exposure to short pulses of very high SO_2 levels. He was able to show an inverse correlation between average chlorophyll content of a species and sensitivity in gassing experiments.

Several workers have noted that the resistance of species to SO_2 has a marked dependence on humidity (Coker 1967). This dependence has been shown to be due in part to the ability of bryophytes to shut down their metabolism in the dehydrated state. Unlike higher plants, many bryophytes also have cell walls that crumple around the shrinking protoplasts thus preventing plasmolysis, and in many species the protoplasts can sustain very high osmotic pressures by shutting down their metabolism until rehydrated. This may be the reason why *Lophocolea heterophylla* and *Hypnum cupressiforme* var. *resupinatum* were so successful at surviving in the bryophyte/lichen desert in the east London area during the worst phases of pollution, and why *Hypnum mammillatum* survives in Epping Forest despite its high sensitivity to SO_2 in the hydrated state (Syratt 1969).

Bryophytes differ considerably in their internal osmotic potentials (OP). Members of the Orthotricaceae, *Cryphaea*, and *Antitrichia* typically have high internal OPs (about 25 atmospheres), making them tolerant of areas of low humidity. On wetting they rehydrate rapidly, however, and are more likely to take in surface-accumulated and dissolved pollutants. More SO_2-tolerant species such as *Mnium hornum* and *Isothecium myosuroides* have lower internal OPs and rehydrate much more slowly (Coker 1967).

DISCUSSION

Relationship between national and local recording

The observations on the effect of atmospheric pollution outlined above illustrate the value of national recording and of detailed local studies, and demonstrate their inter-dependence. The national recording scheme has identified species which appear to have shown very marked responses to atmospheric pollution over large areas. This national picture is, however, inevitably imprecise, as it aggregates records from areas with differing histories of recording and different degrees of recent coverage. It is therefore valuable to be able to test correlations derived from national recording against a dataset derived from London and Essex, one of the few areas with a long history of bryophyte recording.

The lists of epiphytes which are apparently susceptible to atmospheric pollution in Great Britain (Table 2) and Essex (Table 3) were derived separately, by Preston and Adams respectively. (It is impossible to claim, however, that the lists are truly independent as we must share many basic beliefs and assumptions.) There is a good correspondence between the national and local lists: eleven of the 18 species regarded as most susceptible to pollution on the basis of evidence from Essex appear in the national list of adversely affected species, and three of the remainder are suspected of being limited by factors other than SO_2 pollution. Discrepancies between the lists could arise if the evidence of national distribution currently available is insufficient to indicate susceptibility to atmospheric pollution, or if the

decline in Essex was actually caused by other factors.

The effect of atmospheric pollution on the fruiting performance or vegetative vigour of a species is most likely to be identified by intensive local studies. The work of Jones (1952b, 1953) in Berkshire and Oxfordshire, another area with a long history of bryophyte recording, led him to conclude that 19th century specimens of epiphytes such as *Neckera pumila* and *Orthotrichum lyellii* were much more luxuriant than material which he could have gathered in the area. The increase in luxuriance of some mosses which followed the recent decrease in SO_2 pollution in the London area has been discussed above.

Studies in an area which has been as heavily polluted as London also identify species which have declined locally in areas of extremely high SO_2 concentrations, but which are scarcely affected at the national scale. *Grimmia pulvinata, Homalothecium sericeum* and *Orthotrichum diaphanum*, although almost eliminated from central London and the inner suburbs during the period of maximum SO_2 pollution, are nevertheless virtually ubiquitous elsewhere in southern England.

Future recording in a period of falling SO_2 levels

The remarkable return of epiphytic species to the London area described above has been paralleled elsewhere in southern England. In Cambridgeshire, for example, *Ulota crispa* was recorded in 1984 for the first time since 1881; subsequently *Cryphaea heteromalla, Leucodon sciuroides* and *Orthotrichum lyellii* have all been discovered in new sites in the county (Whitehouse 1985; Preston & Whitehouse 1989).

It will be fascinating to observe the sequence in which bryophytes return to the formerly polluted areas. SO_2 levels have fallen so rapidly in London that species are unlikely to recolonise in an order which reflects their sensitivity to pollution. No species are likely to be limited by SO_2 concentrations alone. The sequence of recolonisation is likely to be determined by the proximity of source populations and the mobility of the species. Epiphytes are amongst the more mobile bryophytes. Smith (1982) noted that there is a high proportion of monoecious taxa amongst obligate, as opposed to facultative, epiphytes. He interpreted this finding as an adaptation to promote homozygosity amongst species with specific habitat requirements, but it might also be validly interpreted as an adaptation to maximise the likelihood of sporophyte production, and hence the possibility of colonising a habitat which is regularly created by growing trees. In addition, many epiphytes reproduce asexually by gemmae.

Species may colonise areas from which they have never been recorded in the past. There are, indeed, already signs that this is beginning to happen. Since 1984 *Orthotrichum pulchellum* has been found in five

English vice-counties from which it had not previously been recorded (North Essex, Hertfordshire, West Suffolk, Cambridgeshire and Huntingdonshire). *Ulota phyllantha*, a distinctive epiphyte which is most abundant in coastal areas, has been found in eight new English vice-counties since 1983 (and rediscovered in a further three in which it was believed to be extinct). These records may arise from the recolonisation of areas in which the species previously grew, but was not recorded because of the paucity of 18th and 19th century bryologists. However, habitats which are available to epiphytes now will not be exact replicas of those present before the most severe SO_2 pollution. Eutrophic bark in areas with low SO_2 levels must, for example, be a much commoner habitat now than it was in 1800. It is conceivable that this niche is being exploited by *Ulota phyllantha*, a species known to be tolerant of high ionic concentrations in sea water (Bates & Brown 1974).

Although SO_2 levels have fallen in Britain, it should be remembered that SO_2 levels which were once regarded as innocuous to higher plants nevertheless cause considerable reductions to crop yield under field conditions (Bell 1980). Furthermore, SO_2 pollution is still a major problem on the Continent of Europe and in North America. An accurate bryophyte scale comparable to that for lichens (Hawksworth & Rose 1970) would be a valuable additional tool for the bioassay of pollution levels. The continuing rise in the atmospheric levels of nitrogen oxides, ozone, ammonia and an increasing number of organic compounds also emphasises the importance of a detailed baseline knowledge of bryophyte distribution which can be used if one of these pollutants reaches critical levels.

NOTES

1. The bibliographies we have compared are *Literature on air pollution and lichens* XII–XXX (Henderson 1980–89) and the references listed under the heading 'Pollution' in *Recent bryological literature 53–72* (Clarke 1980–89). Henderson's bibliography includes any reference of relevance to the study of air pollution and lichens, whereas Clarke only lists papers under the heading 'Pollution' if they deal predominantly or exclusively with this topic. Additional bryological references which are relevant to pollution studies will appear under other headings.

2. Grindon's *Manchester walks and wild-flowers* is undated. According to the preface, it was based on articles published in the *Manchester Weekly Times* from May to July 1858. Simpson (1960) and Desmond (1977) give the date of publication as 1858. Internal evidence strongly suggests that 1859 was the true date of publication. On p109, Grindon contrasts the early spring vegetation at Mobberley and Heywood 'this present season (1859)'. The title page refers to Grindon as 'author of *The Manchester flora*', which was published in 1859. In the preface to *The Manchester flora*, dated 28 March 1859, Grindon writes of 'its little companion, *Manchester walks and wild-flowers*, published simultaneously'.

REFERENCES

Adams, K.J. 1990. Proposals for a 5-km^2 mapping scheme for eastern England. *Bulletin of the British Bryological Society*, **55**, 14–17.

Bates, J.W. & Brown, D.H. 1974. The control of cation levels in seashore and inland mosses. *New Phytologist*, **73**, 483–495.

Bell, J.N.B. 1980. Response of plants to sulphur dioxide. *Nature, London*, **284**, 399–400.

Bowen, H.J.M. 1968. *The flora of Berkshire*. Privately published.

Bowen, H.J.M. 1980. A lichen flora of Berkshire, Buckinghamshire and Oxfordshire. *Lichenologist*, **12**, 199–237.

Briggs, A. 1968. *Victorian cities*, 2nd ed. London: Pelican Books.

Burton, M.A.S. 1986. *Biological monitoring of environmental contaminants (plants)*. (MARC Report no. 32.) London: Monitoring and Assessment Research Centre.

Chandler, T.J. 1976. The climate of towns. In: *The climate of the British Isles*, edited by T.J. Chandler & S. Gregory, 307–329. London: Longman.

Clarke, G.C.S. 1980–89. Recent bryological literature 53–72. *Journal of Bryology*, **11**(1)–**15**(4).

Coker, P.D. 1967. The effects of sulphur dioxide pollution on bark epiphytes. *Transactions of the British Bryological Society*, **5**, 341–347.

Coppins, B.J. 1973. The 'drought hypothesis'. In: *Air pollution and lichens*, edited by B.W. Ferry, M.S. Baddeley & D.L. Hawksworth, 124–142. London: Athlone Press.

Crombie, J.M. 1885. On the lichen-flora of Epping Forest, and the causes affecting its recent distribution. *Transactions of the Essex Field Club*, **4**, 54–75.

Department of the Environment. 1980. *Air pollution in the Bedfordshire brickfields*. London: Department of the Environment.

Desmond, R. 1977. *Dictionary of British and Irish botanists and horticulturists*, 272. London: Taylor & Francis.

Dickson, J.H. 1973. *Bryophytes of the Pleistocene*. Cambridge: Cambridge University Press.

Dickson, J.H. 1981. Mosses from a Roman well at Abingdon. *Journal of Bryology*, **11**, 559–560.

Disraeli, B. 1844. *Coningsby; or, the new generation*. 3 vols. London: Henry Colburn.

Ferguson, P. & Lee, J.A. 1980. Some effects of bisulphite and sulphate on the growth of *Sphagnum* species in the field. *Environmental Pollution*, (A), **21**, 59–71.

Ferguson, P. & Lee, J.A. 1983. The growth of *Sphagnum* species in the southern Pennines. *Journal of Bryology*, **12**, 579–586.

Ferguson, P., Lee, J.A. & Bell, J.N.B. 1978. Effects of sulphur pollutants on the growth of *Sphagnum* species. *Environmental Pollution*, **16**, 151–162.

Ferguson, P., Robinson, R.N., Press, M.C. & Lee, J.A. 1984. Element concentrations in five *Sphagnum* species in relation to atmospheric pollution. *Journal of Bryology*, **13**, 107–114.

Gilbert, O.L. 1968. Bryophytes as indicators of air pollution in the Tyne valley. *New Phytologist*, **67**, 15–30.

Gilbert, O.L. 1971. Urban bryophyte communities in north-east England. *Transactions of the British Bryological Society*, **6**, 306–316.

Grindon, L.H. 1859a. *Manchester walks and wild-flowers*, 1. London: Whittaker.

Grindon, L.H. 1859b. *The Manchester flora*, 513. London: W. White.

Harrop, O., Laxen, D. & Daunton, R. 1985. Sulphur dioxide in London's air. *London Environmental Bulletin*, **2**(4), 15.

Hawksworth, D.L. & McManus, P.M. 1989. Lichen recolonization in London under conditions of rapidly falling sulphur dioxide levels, and the concept of zone skipping. *Botanical Journal of the Linnean Society*, **100**, 99–109.

Hawksworth, D.L. & Rose, F. 1970. Qualitative scale for estimating sulphur dioxide air pollution in England and Wales using epiphytic lichens. *Nature, London*, **227**, 145–148.

Henderson, A. 1980-89. Literature on air pollution and lichens, XII–XXX. *Lichenologist*, **12**(1)–**21**(4).

Hill, M.O. 1988. A bryophyte flora of north Wales. *Journal of Bryology*, **15**, 377–491.

Hill, M.O., Preston, C.D. & Smith, A.J.E., eds. 1991. *Atlas of bryophytes of Britain and Ireland; Vol. 1, Liverworts (Hepaticae and Anthocerotae)*. Colchester: Harley Books.

James, P.W. 1973. Introduction. In: *Air pollution and lichens*, edited by B.W. Ferry, M.S. Baddeley & D.L. Hawksworth, 1–5. London: Athlone Press.

Jermyn, S.T. 1974. *Flora of Essex*. Colchester: Essex Naturalists' Trust.

Jones, E.W. 1952a. Some observations on the lichen flora of tree boles, with special reference to the effect of smoke. *Revue bryologique et lichénologique*, **21**, 96–115.

Jones, E.W. 1952b. A bryophyte flora of Berkshire and Oxfordshire. I. Hepaticae and *Sphagna. Transactions of the British Bryological Society*, **2**, 19–50.

Jones, E.W. 1953. A bryophyte flora of Berkshire and Oxfordshire. II. Musci. *Transactions of the British Bryological Society*, **2**, 220–277.

Lawther, P.J. 1965. Air pollution. *Discovery, London*, **26** (12), 14–18.

LeBlanc, F. 1961. Influence de l'atmosphere polluée des grandes agglomérations urbaines sur les épiphytes corticoles. *Revue Canadienne de Biologie*, **20**, 823–827.

LeBlanc, F. & Rao, D.N. 1973a. Evaluation of the pollution and drought hypotheses in relation to lichens and bryophytes in urban environments. *Bryologist*, **76**, 1–19.

LeBlanc, F. & Rao, D.N. 1973b. Effects of sulphur dioxide on lichen and moss transplants. *Ecology*, **54**, 612–617.

LeBlanc, F. & Rao, D.N. 1974. A review of the literature on bryophytes with respect to air pollution. *Bulletin Société Botanique de France. Colloque*, **121**, 237–255.

LeBlanc, F. & De Sloover, J. 1970. Relations between industrialization and the distribution and growth of epiphytic lichens and mosses in Montreal. *Canadian Journal of Botany*, **48**, 1485–1496.

LeBlanc, F., Comeau, G. & Rao, D.N. 1971. Fluoride injury symptoms in epiphytic lichens and mosses. *Canadian Journal of Botany*, **49**, 1691–1698.

Lee, J.A. 1986. Nitrogen as an ecological factor in bryophyte communities. [Abstract.] *Bulletin of the British Bryological Society*, **47**, 6–7.

Munday, P.K. 1990. *UK emissions of air pollutants 1970–1988*. (Warren Spring Laboratory Report LR764 (AP) M.) Stevenage: Department of Trade and Industry.

Pentecost, A. 1987. The lichen flora of Gwynedd. *Lichenologist*, **19**, 97–166.

Press, M., Ferguson, P. & Lee, J. 1983. 200 years of acid rain. *Naturalist, Hull*, **108**, 125–129.

Press, M.C., Woodin, S.J. & Lee, J.A. 1986. The potential importance of an increased atmospheric nitrogen supply to the growth of ombrotrophic *Sphagnum* species. *New Phytologist*, **103**, 45–55.

Preston, C.D. & Whitehouse, H.L.K. 1989. Bryophyte records. *Nature in Cambridgeshire*, **31**, 65–66.

Rao, D.N. & LeBlanc, F. 1967. Influence of an iron-sintering plant on corticolous epiphytes in Wawa, Ontario. *Bryologist*, **70**, 141–157.

Rose, C.I. & Hawksworth, D.L. 1981. Lichen recolonization in London's cleaner air. *Nature, London*, **289**, 289–292.

Rose, F. 1974. The epiphytes of oak. In: *The British oak*, edited by M.G. Morris & F.H. Perring, 250–273. Faringdon: Classey.

Rose, F. & Wallace, E.C. 1974. Changes in the bryophyte flora of Britain. In: *The changing flora and fauna of Britain*, edited by D.L. Hawksworth, 27–46. (Systematics Association Special Volume, no. 6.) London: Academic Press.

Rydzak, J. 1959. Influence of small towns on the lichen vegetation. VII. Discussion and general conclusions. *Annales Universitatis Mariae Curie-Sklodowska C*, **13**, 275–323.

Saunders, P.J.W. & Wood, C.M. 1973. Sulphur dioxide in the environment: its production, dispersal and fate. In: *Air pollution and lichens*, edited by B.W. Ferry, M.S. Baddeley & D.L. Hawksworth, 6–37. London: Athlone Press.

Schumacker, R., ed. 1985. *Atlas de distribution des bryophytes de Belgique, du Grand-Duché de Luxembourg et des régions limitrophes. I. Anthocerotae & Hepaticae (1830–1984)*. Meise: Jardin botanique national de Belgique.

Simpson, N.D. 1960. *A bibliographical index of the British flora*, 176. Bournemouth. Privately published.

Skye, E. 1958. Luftföroreningars inverkan på a busk-och bladlavfloran kring skifferoljeverket in Närkes kvarntop. *Svensk Botanisk Tidskrift*, **52**, 133–190.

Smith, A.J.E., ed. 1978. *Provisional atlas of the bryophytes of the British Isles*. Huntingdon: Biological Records Centre.

Smith, A.J.E. 1982. Epiphytes and epiliths. In: *Bryophyte ecology*, edited by A.J.E. Smith, 191–227. London: Chapman & Hall.

Stevenson, R. 1986. Bryophytes from an archaeological site in Suffolk. *Journal of Bryology*, **14**, 182–184.

Syratt, W.J. 1969. *The effect of atmospheric pollution on epiphytic bryophytes*. PhD thesis, University of London.

Thomas, M.D. 1961. *Effects of air pollution on plants*. In: *Air pollution*. (WHO Monograph 46.) Geneva: World Health Organisation.

Touw, A. & Rubers, W.V. 1989. *De Nederlandse bladmossen*. Utrecht: Stichting Uitgeverij van de Koninkklijke Nederlandse Natuurhistorische Vereniging.

Wallace, E.C. 1963. Distribution maps of bryophytes in Britain. *Ptilidium pulcherrimum* (Weber) Hampe. *Transactions of the British Bryological Society*, **4**, 513.

Warren Spring Laboratory. 1969-72. *The investigation of air pollution. National survey of smoke and sulphur dioxide. Directory and tables*. London: Department of Trade and Industry.

Warren Spring Laboratory. 1972. *National survey of air pollution 1961–71*, 1. London: HMSO.

Warren Spring Laboratory. 1979-88. *UK smoke and sulphur dioxide monitoring networks. Summary tables*. Stevenage: Department of Trade and Industry.

Whitehouse, H.L.K. 1985. Bryophyte records. *Nature in Cambridgeshire*, **27**, 7–8.

Winner, W.E. 1988. Responses of bryophytes to air pollution. *Bibliotheca Lichenologica*, **30**, 141–173.

Winner, W.E., Atkinson, C.J. & Nash, T.H. 1988. Comparison of SO_2 absorption capacities of mosses, lichens and vascular plants in diverse habitats. *Bibliotheca Lichenologica*, **30**, 217–230.

Introductions and their place in British wildlife

B C Eversham and H R Arnold

Biological Records Centre, Environmental Information Centre, Institute of Terrestrial Ecology, Monks Wood Experimental Station, Abbots Ripton, Huntingdon, Cambs PE17 2LS

The topic of introductions and invasions is a wide one, and has been the subject of much debate among ecologists in recent years, with a Royal Society discussion meeting on biological invasions in 1986 (Kornberg & Williamson 1987), a British Ecological Society meeting on colonisation, succession and stability (Gray, Crawley & Edwards 1987), which also paid close attention to the impact of invaders, and a Mammal Society/Fauna and Flora Preservation Society symposium on reintroductions of mammals in 1986 (Anon 1986).

Some of the most dramatic recent incidents of long-range introductions are in Africa, and, while not having space to examine them in detail, they serve as a reminder that we have in no sense 'learnt our lesson' from ecological disasters at home and abroad. Slipshod quarantine procedures, inadequate fumigation of aircraft, and careless pet and livestock keepers are as prevalent as ever. The massive increase in global travel and the speed of commerce – fresh *mangetout* with fresh long-tailed blue butterfly (*Lampedes boeticus*) caterpillars in our supermarkets every week – mean that the opportunist plant or animal can reach Britain more easily than ever before.

Concentrating on impacts on *wildlife* neglects the problems that agriculture and forestry face from aliens. Most pests of crops in Britain are not native species, and British species have often become serious agricultural pests when introduced elsewhere in the world (eg molluscs in Australia (Baker 1989), New Zealand (Barker 1989) and USA (Barrett, Byers & Bierlein 1989)); this is a signpost to an underlying pattern to be discussed later.

Excluding the above topics, what remain are the diverse impacts of introduced species as environmental factors on the native British flora and fauna.

Crucial to the discussion of introductions is Britain's island status. Many of the species that we regard as introductions reached the adjacent Continent by their own efforts several thousand years ago. It was only the English Channel and the North Sea, a peculiarity of the present inter-glacial, that necessitated human assistance for their transport to Britain.

To quantify the effect of being an island, the native fauna and flora of Britain and of Ireland can be compared with that of an equal area of the adjacent

Table 1. Numbers of species in selected animal groups native in Ireland, Britain, an equivalent Continental area (defined in Figure 1) and western Europe (where figures are available). Data from distribution maps in the following publications: Corbet and Ovenden (1980), Stebbings (1988), Peterson, Mountford and Hollom (1983), Arnold and Burton (1978), Maitland (1977), Kerney and Cameron (1979), Higgins and Riley (1983)

Group	Ireland	Britain	Continental equivalent area	W Europe
Quadruped mammals	11	26	42	103
Bats	6	14	22	30
Breeding birds	126	186	221	347
Reptiles	1	6	15	49
Amphibians	3	6	16	39
Freshwater fish	20	31	42	153
Land molluscs	66	95	147	*
Butterflies	29	62	126	349

* Data not available

Table 2. Numbers of species in selected vascular plant groups native in Ireland, Britain, an equivalent continental area (defined in Figure 1) and western Europe. Data from *Atlas Florae Europaeae* (Jalas & Suominen 1972–86)

Group	Ireland	Britain	Continental equivalent area	W Europe
Pteridophyta	55	66	70	142
Gymnospermae	2	5	4	38
Ranunculaceae	27	38	55	291
Caryophyllaceae (Sandworts, etc)	28	53	60	308
Caryophyllaceae (Campions, etc)	7	19	31	404
Other Centrospermae	18	27	37	103
Amentiferae (*Salix*, etc)	23	35	38	125
Polygonaceae	20	28	30	77

Continental mainland. Figure 1 shows such an area, of approximately the same areal extent as Britain and Ireland, based on the 50 km squares of the Universal Transverse Mercator Grid, and lying closest to southeast England (the most likely point of entry of terrestrial post-glacial colonists). Tables 1 and 2 illustrate the paucity of the British fauna and flora compared with the Continental mainland (both the 'equivalent area' of Figure 1, and western Europe as a whole), and the further reduction in diversity in Ireland, due at least in part to an additional post-glacial sea barrier (Godwin 1975).

Being an island has also shaped the history of Britain: 'after the discovery of America and the ocean routes to Africa and the East, Britain lay at the centre of the new maritime movement' (Trevelyan 1926). A maritime nation was more inclined to exploration and trade by sea (Morrill 1988) and, in consequence, Britain probably received its first transatlantic immigrants sooner than most other European countries. The range and diverse origins of garden plants today is breathtaking; but at least as amazing

is how many of them had reached Britain by the 18th century.

By contrast with the British Isles, a few island groups are noted for their endemic biotic diversity, and have suffered catastrophically from the impact of introductions. These are *oceanic* islands, such as Hawaii, New Zealand and Galapagos, where species have been isolated for long enough to have speciated, and in some cases radiated extensively. As a very recent *continental* island, Britain has few endemics, and almost all of them are the final remnant of a formerly much wider geographic range.

A feature of oceanic islands which permits a high incidence of endemism is the absence of mammalian predators and scavengers. Most island-endemic birds have evolved without the pressures of rodents, cats and mustelids, and thus are very vulnerable to their introduction. There appear to be conspicuous 'empty niches' which these and other invaders such as feral goats (*Capra hircus*) are able to fill, or niches occupied by 'inept' animals: birds or reptiles

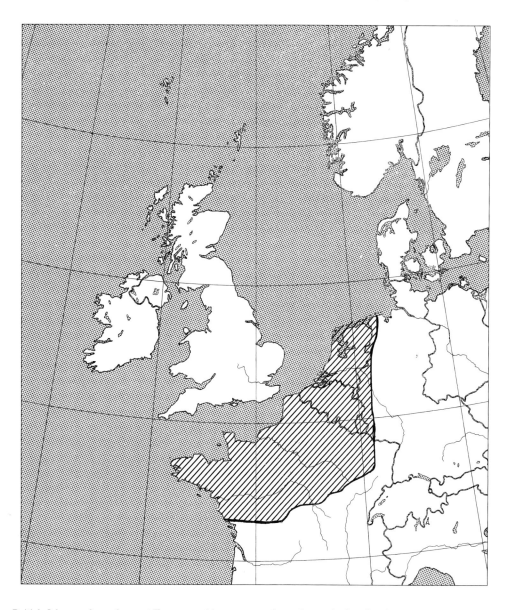

Figure 1. The British Isles and north-west Europe, with an approximately equivalent land area on the Continent adjacent to Britain shaded; this is the area used in the compilation of Tables 1 and 2

occupying niches usually associated with mammals are easily out-competed.

To put Britain's many aliens into a broader context: the native Hawaiian avifauna can be accounted for by only one successful colonist every 350 000 years (Elton 1958; Mollison 1986). Similarly, the large numbers of endemic species on islands in the Mediterranean (such as *c* 160 flowering plants on Crete, 10% of the native flora (Turrill 1929; Polunin 1980)) or off North Africa (blue chaffinch (*Fringilla teydea*) and endemic chat (*Saxicola dacotiae*) on the Canary Isles, 16 species of snails (Collins & Wells 1987) and 25 species of *Cylindroiulus* millipedes (Enghoff 1982) on Madeira) are all indicative of a long and stable history. The main difference in this case is latitude: the islands are far enough south to have escaped the many glaciations which repeatedly erased the temperate fauna and flora from the land which is now Britain.

WHAT IS 'NATIVE'?

Most definitions of 'native' simply mean that a species arrived unaided by man. Unfortunately, that is very hard to prove, unless the arrival itself is witnessed. Webb (1985) discussed the criteria for assuming native status for vascular plants and clarified the definition. To be native, a species must either have arrived here before the Neolithic period (*c* 5000 BP (before present)), when man first began farming here, or have arrived since then completely independent of man's activities. An alien is thus a species whose arrival is a consequence of the activities of man or his domestic animals.

An additional qualifier would be that species should successfully reproduce in Britain. Regular non-established migrants, such as clouded yellow butterflies (*Colias* spp.), and non-breeding residents, such as the mouse-eared bat (*Myotis myotis*) in Sussex or the black-browed albatross (*Diomedea melanophris*) in the Shetlands, are barely natives. As the individuals involved are unlikely to play any part in producing future generations, they are of minimal conservation value except as publicity, but they could help to form the nucleus of an establishing population, or function as a 'honeypot' for colonial species.

Wintering birds are a special case. For many species, Britain plays a vital role as the main feeding area, maintaining the condition of a population, outside the breeding season. Despite this importance, they will not be considered further in this paper, although it is possible for winterers to become resident, eg occasional fieldfares (*Turdus pilaris*) and bramblings (*Fringilla montifringilla*) (Sharrock 1976), or twite (*Carduelis flavirostris*) (Marshall, Lynes & Limbert 1989), in the lowlands.

A change in climate could, of course, alter the status of many of these migrant species. Last century, it seems the continental race, subspecies *gorganus*, of the swallowtail butterfly (*Papilio machaon*) was estab-lished in Kent and adjoining counties for several successive years, but at present it is only an occasional vagrant (Bretherton 1989). Similarly, there are a few vagrant *Sympetrum* dragonflies which have bred in southern England, but do not do so regularly (Merritt, Moore & Eversham 1992).

Webb (1985) recognised that, in many cases, it will be impossible to decide whether a plant was introduced as a weed of neolithic cultivation, or was already established in Britain and appears in the pollen record at that time because it became more abundant in the early open fields.

Presence in previous inter-glacials is not evidence that a species is native now: *Rhododendron ponticum* reached Ireland in the Hoxnian or Gortian inter-glacial *c* 150 000 years ago (Godwin 1975), but there are no more recent fossil records, and no native occurrences of it in northern Europe since the last glaciation. Its present natural distribution is disjunct, very local in southern Spain and Portugal, and rather more widespread around the Black Sea (Tutin *et al.* 1972). Contrary to many accounts, *R. ponticum* is not a Himalayan species. Surprisingly, there seem to be some doubts as to whether the vigorous and widespread rhododendron in Britain is pure *R. ponticum* or a hybrid involving *R. catawbiense*, native in the Appalachian Mountains, eastern USA (Cross 1975). The presence of rhododendrons in Britain now is because of deliberate introduction in the 18th century.

Post-glacial fossil evidence that a species was established in Britain between the end of the last glaciation and the beginning of agriculture is almost conclusive proof of native status. The exceptions will be species which have become extinct here, and been reintroduced, such as the red squirrel (*Sciurus vulgaris*) in Scotland (Harvie-Brown 1880–81) and in Ireland (Barrington 1880), or the Scots pine (*Pinus sylvestris*) in Ireland and perhaps England (Huntley & Birks 1983; Clapham, Tutin & Moore 1987).

The very early documentation of a few species must nevertheless be treated with caution. For example, the fallow deer (*Dama dama*) is frequently described or figured in medieval manuscripts as if it were a well-known resident, yet it had been established in Britain for only a few centuries; and woad (*Isatis tinctoria*) features in Roman histories of the Celts, although it is believed to be native to southern and eastern Europe (Jessen & Halbaek 1944; Godwin 1975). In the absence of fossil evidence, a species must be shown to have arrived without human aid, which is obviously impossible to prove in most cases.

Genetic divergence: if, like caper spurge (*Euphorbia lathyrus*), a species is believed to be native in some places, but an introduction in others, there might be genetic differences between populations. The native fenland swallowtail butterfly (*Papilio machaon*) differs significantly from the continental race (Hall & Emmet 1989); and there is some evidence that sticky groundsel (*Senecio viscosus*) in natural habitats such as shingle beaches differs from populations on waste ground and railway lines (Akeroyd, Warwick & Briggs 1978). Unfortunately, very few studies of this kind have been performed, and it is a costly and time-consuming activity. It is also possible that apparent divergence can arise through founder-effects and genetic drift in a remarkably short time (eg Berry's work on house mice (*Mus musculus*) (Berry & Jakobson 1974)).

The study of enzyme polymorphisms has provided a further useful if laborious technique, which has established that some species of molluscs in the genus *Arion* are represented in Britain both by an out-crossing (long-established, probably native) and an obligately self-fertilizing (recently introduced) strain (Foltz *et al*. 1982); the same alien strains have been detected in the USA, in the absence of their sexual counterpart.

Habitat: most native species occur in 'natural' habitats, at least occasionally – sand dunes support many of the native weeds of agricultural land. Many aliens are confined to man-made sites, eg the woodlouse *Porcellionides pruinosus* in dung heaps and compost (Harding & Sutton 1985; Sutton & Harding 1989); and any species which is usually found at sites with a high proportion of certain aliens must be 'doubtful'.

Geographic distribution: an isolated population of a species hundreds of miles from the main range may be considered doubtfully native. There are exceptions, some of which can be justified as the last vestige of a once-wider range. For example, many insects associated with ancient woodland are now restricted to isolated sites (Harding 1978), or even to individual trees as in the case of *Limoniscus violaceus* (Welch 1987), some of which form a pattern. If several species show the same disjunction, it may have a natural explanation. Most species with relict distributions will occur at several scattered sites, and it may be possible to identify the features of history and management that explain their survival.

Historical evidence of introduction can be conclusive, as in the case of *Buddleia davidii*, which first appeared on sale as a garden plant in 1896 (Webb 1985). Often, however, the historical information may apply to only some of a species' populations in Britain, as perhaps in the case of caper spurge referred to earlier, which is often grown in gardens, yet has been used as an indicator of ancient woodland in eastern England (Rixon & Peterken 1975).

Rapid declines or expansions are often, though not always, symptoms of alien status. Recent colonists such as American willowherb (*Epilobium ciliatum*) obviously expand their range while becoming established. Some, such as the famous Canadian pondweed (*Elodea canadensis*), undergo an initial flush then a slower reduction in abundance. It is worth noting that aliens may represent taxonomic problems: while New Zealand willowherb (*Epilobium brunnescens*) was recognised instantly (there being no prostrate *Epilobium* in Europe), *E. ciliatum* was confused with several natives, and overlooked, as was *Elodea nuttallii*. Some aliens, such as the corn-cockle (*Agrostemma githago*), have declined spectacularly.

Frequency of naturalisation: if a species is known to be introduced at many of its sites, the status of populations whose origin is unknown may be suspected of introduction. This applies to many liliaceous plants which are often cultivated, such as grape-hyacinth (*Muscari atlanticum*) and lily-of-the-valley (*Convallaria majalis*). At certain sites, however, they are believed to be native.

Inability to reproduce: plants which cannot set seed in Britain might be suspected to have been introduced. Thus, the ubiquitous horseradish (*Armoracia rusticana*) which rarely produces ripe fruit in Britain (Clapham *et al*. 1987) is, not surprisingly, an alien (though how it and other species become so very widespread with no obvious means of long-distance dispersal is a puzzle). Likewise, the failure of some species to breed in part of their range, such as large-leaved lime (*Tilia platyphyllos*) in Scotland (Pigott 1981), implies at least local translocation. Zoological examples are fewer, mainly because animals tend to be less long-lived than perennial plants. Migrant Lepidoptera, such as the clouded yellow butterfly (*Colias* spp.) and the silver Y moth (*Autographa gamma*), can be seen in areas well beyond their breeding range. A number of species of ant of the genus *Camptonotus* have been introduced in imported timber, and survive for a short period in the vicinity of timber yards, but have not so far become established (Bolton & Collingwood 1975).

Means of introduction: if some or all of the known populations of a species are close to sources of introduced material, a strong case would be needed to prove native status. Occasionally, such circumstantial evidence can be misleading: the earliest records of the mouse-eared bat (*Myotis myotis*) were from the British Museum, Bloomsbury, prior to 1850, and from the grounds of Girton College, Cambridge, in 1888. Both are sites where one might expect to find imported animals, but equally are places where one could expect to find people capable of identifying unusual species!

All these criteria are tentative rather than absolute proof, and many indicate possible introductions rather than possible natives. They can lead one to doubt the status of a species unjustly. For instance, two beetles, *Curimopsis nigrita* and *Bembidion humerale*, were first found in Britain at a bog in South Yorkshire (Crossley & Norris 1975; Johnson 1978). For several years, they were known from no other site. Their European distribution is centred on the Baltic (ie the Yorkshire site is an extreme outlier). The site is a cut-over raised mire (arguably a man-modified habitat). At the turn of the century, a number of barges were imported from Holland (where both the species occur) for use in transporting the cut peat. There was thus considerable evidence to

Table 3. Bird species which have expanded their range and/or increased in abundance in the British Isles since 1700, and the approximate time of the increase. + indicates species which have also increased in other parts of their range. Based on information in Sharrock (1976), Cramp (1977, 1980, 1983, 1985, 1988). Fisher (1966) and Parslow (1973)

Species	Approx. dates of expansion	European expansion
Great crested grebe	1870–present	+
Fulmar	1750–present	+
Gannet	1900–present	?
Goosander	1870–present	
Oystercatcher	1900–present	
Little ringed plover	1950–present	+
Woodcock	1890–1920	
Curlew	1910–1960	+
Great skua	1890–present	
Great black-backed gull	1880–present	+
Lesser black-backed gull	1900–present	+
Herring gull	1900–present	+
Common gull	1870–present	+
Black-headed gull	1900–1980	+
Kittiwake	1900–present	+
Stock dove	1820–1950, 1965–present	+
Woodpigeon	1820–present	+
Turtle dove	1820–present	
Green woodpecker	1820–present	+
Great spotted woodpecker	1870–present	+
Golden oriole	1960–?	+
Jay	1910	?
Mistle thrush	1750–present	+
Fieldfare	1960–present	+
Black redstart	1920–present	+
Wood warbler	1850–?	?
Firecrest	1950?–present	+
Pied flycatcher	1940–?	?
Starling	1830–present	?
Siskin	1850–present	?
Serin	1960–?	+

imply possible introduction. Since then, the two species have now been found at another site which has no known Dutch connections (Skidmore, Limbert & Eversham 1987). *Curimopsis nigrita* has since been identified as a Bronze Age fossil (Buckland 1979; Buckland & Johnson 1983), which is certain proof of native status, as it is not a synanthrope.

'NATURAL' INVASIONS

In 1952, the collared dove (*Streptopelia decaocta*) was first recorded in Britain (May & Fisher 1953). In view of its 'dramatic and unprecedented spread south-west across Europe' in the previous 20 years, this individual bird was regarded by the authors as a 'pioneer of the spread rather than an escape from captivity'. Editorial comment, however, said 'Meanwhile we feel bound to conclude that no adequate evidence has so far been produced for adding

S. decaocta to the British list', but in the previous sentence remarked that 'further and less controversial occurrences will soon follow', which proved to be true. The species was first recorded breeding in 1955, and has now colonised most of Britain and Ireland (Sharrock 1976). Details of its spread across Europe are described by Hengeveld (1989).

Many other British or European birds are expanding or have recently expanded significantly. Table 3 lists some of the more dramatic examples, and indicates which have expanded over other parts of their range as well as in the British Isles. (Species whose expansion is solely due to the abatement of direct human pressures, such as birds of prey harassed by gamekeepers or poisoned by pesticides, are not listed.)

One is left to ponder whether such striking changes are therefore merely a slightly more extreme example of the way most species behave in the post-glacial north temperate zone.

TRANSLOCATIONS WITHIN THE BRITISH ISLES, AND REINTRODUCTIONS

Reintroductions and translocations for the purpose of 'topping up' declining or low populations have taken place within the British Isles. In the 17th century, the red squirrel had become extremely rare or extinct over much of Scotland, due mainly to the destruction of the forests. In the 18th and 19th centuries, many reintroductions took place using stock from England (eg at Dalkeith in about 1772) and from within Scotland (eg at Minto, Roxburgh, from Dalkeith, in 1827) (Harvie-Brown 1880–81; Lever 1977). There also seems to have been a natural increase in range from the remnants of the old forest areas, into new plantations, during this period. In addition, some squirrels were released which apparently came from the Continent.

Many amateur naturalists breed native butterflies, and some release surplus adults. There has been a great increase in interest in 'wildflower meadow seed mixtures' in recent years (Wells, Bell & Frost 1981), and, although most are probably sown in gardens or in sites such as new road embankments, some have been sown in nature reserves. The motives for such activities are complex – almost always for the best of reasons, but not always with sufficient forethought for the possible consequences. The biggest problem is inadequate documentation and lack of consultation with those likely to be interested or affected (eg nature reserve managers). Are such activities really 'good conservation', or are they the naturalists' equivalent of rearranging the deck chairs on the Titanic?

The success or failure of attempts at translocation of once-native or currently rare species depends largely on careful planning and a thorough knowledge of a species' ecological needs. The reintroduction of the large blue butterfly (*Maculinea arion*) can be seen as the culmination of years of research, originally intended to safeguard the native population, but started a few years too late. Reintroduction may be reasonable when all else has been thoroughly tried. Perhaps the public or political pressure to produce a 'positive' result in such cases makes an attempt almost inevitable.

The marsh fritillary (*Eurodryas aurinia*) is known to have disappeared from most of eastern England, largely through the drainage or improvement of old pasture (Heath, Pollard & Thomas 1984), yet a few sites exist which are now, as nature reserves, in perfect condition for this species. It is presumed that it cannot recolonise these sites naturally, so there is a choice. Some reintroductions, such as natterjack toads (*Bufo calamita*) at Sandy, Bedfordshire, are apparently outstanding successes.

One hopes that conservationists have learnt their lesson, and attempts to introduce species in areas completely devoid of suitable habitats or climate, like the release of sand lizards in the Hebrides (Lever 1977), will not be repeated: native stocks are no longer 'buoyant' enough to withstand regular 'harvesting' for translocation.

WHY ARE SPECIES INTRODUCED?

Accidents of cultivation and commerce

Sweet cicely (*Myrrhis odorata*) may be native, but has been widely grown as a culinary herb (it has a strong aniseed flavour, and was used to flavour brandy, and to mask the flavour of unsavoury meats in the days before refrigerators (Mabberley 1987)). Deadly nightshade (*Atropa belladonna*) is less easy to explain: it may have been a medicinal herb, or grown for its showy berries.

Various European amphibians have been found at or near Beam Bridge nurseries in Surrey, a horticultural nursery selling aquatic plants. Animals could have been transported as adults, larvae or as eggs amongst bundles of pond weed. Deliberate introductions of the European tree frog (*Hyla arborea*) at several sites have been unsuccessful. The reason may be that the donor populations on the Continent were almost all male: females visit ponds only briefly, to lay eggs, but males linger all summer, so are far more likely to be collected by herpetologists. It appears there is now a population established on the Isle of Wight; this species may benefit from any future climatic warming.

The brown rat (*Rattus norvegicus*) seems to have been introduced in the 18th century – 1728 or 1729 are dates widely quoted – and it probably first arrived on board ships from Russian ports (Barrett-Hamilton & Hinton 1910–21; Twigg 1975).

House mice are often transported in food, bales of straw, and other cargo. The history of island populations of this and other small rodents has been investigated in detail by Berry (1963, 1968, 1970), who has shown that significant genetic differences between island and mainland populations can develop within a few decades. The apparent accidental transport of hedgehogs (*Erinaceus europaeus*) to off-shore islands as stowaways among plant material is discussed by Morris (1983). Hedgehogs may, on occasion, have been introduced to off-shore islands deliberately, either to control 'pests' in gardens, or for sentimental reasons. They are now perceived by many conservationists as a potential threat to ground-nesting seabirds, for which many islands are noted; as such, their translocation is to be discouraged.

There is a long list of casual records of non-British species from the vicinity of ports, such as Newcastle-upon-Tyne, but few have successfully become established. The clearest case is probably the large carabid, *Pterostichus cristatus*. This species is abundant in north-east England, and has the appearance of a native (Luff 1982): it is the commonest carabid in many riverside woods, replacing the certainly native *Pterostichus madidus* and *P. niger* locally.

Deliberate introductions

Economic motives

Rabbits (*Oryctolagus cuniculus*) were originally bred for their skins and their meat, and often proved very difficult to establish, even when they were looked after with great care! Their history has been thoroughly documented by Sheail (1971). Pheasants (*Phasianus colchicus*) may similarly have been introduced for food, probably in the 11th century; they were quite widely naturalised in the 12th century (Lever 1977).

The coypu (*Myocastor coypus*) was imported into fur farms in the 1930s, for its pelt, known as 'nutria'. Inevitably, escapes occurred, especially when the Second World War led to a lack of maintenance of perimeter fences. Up to the mid-1940s (Figure 2), there were many scattered records, but only two areas where coypu bred – in Berkshire and in the Norfolk Broads. The Broads area became the centre for a considerable population explosion, even though numbers were greatly reduced by the cold winter of 1947–48. A wetland species, the coypu burrowed into the banks of ditches and fed voraciously on reeds, and on sugar beet and other crops. In a low-lying county such as Norfolk, with much of the land close to sea level, the threat of impeded drainage was even greater than the direct losses to agriculture (Figure 3). The Ministry of Agriculture, Fisheries and Food responded with the Coypu Control Campaign in 1962. This campaign helped reduce the size of the population, and coincided with another severe winter, 1962–63; but numbers appeared to increase again in the 1970s. A further concerted trapping effort began then and continued throughout the 1980s, so that in 1989 only two were recorded (Figure 4). If the eradication of coypu has been successful, it has probably eliminated another, accidental, introduction – the host-specific parasitic louse *Pitrufquenia coypus*.

During peaks of population size, coypu exerted a

marked effect on semi-natural vegetation. In particular, coypu numbers have been linked with the large-scale dieback of reedswamp (Boorman & Fuller 1981). Unfortunately, there is little sign of a recovery of reedswamp since the demise of the coypu. Since the 1940s, other changes, such as pollution, eutrophication and increased boating traffic, have all affected the Broads.

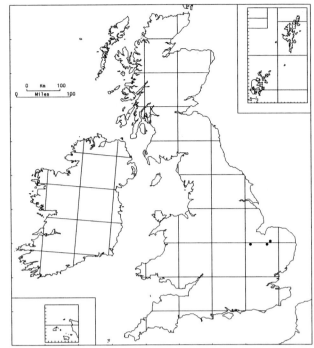

● 1988–1990

Figure 4. Records of coypu in Britain, 1988–90

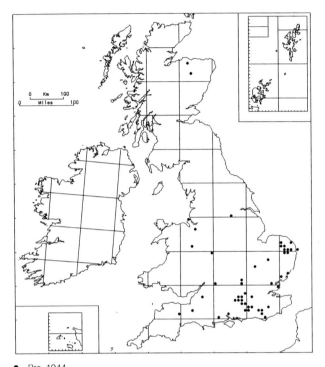

● Pre–1944

Figure 2. Records of coypu in Britain up to 1944

Two other escapees from fur farms appear to have the potential for significant ecological impact. One, the muskrat (*Ondatra zibethicus*), has so far failed to establish itself permanently in Britain, although it has colonised parts of the adjacent Continent very successfully (Lever 1977; Gosling & Baker 1989). Feral populations in both Britain and Ireland, which established in the 1930s, were exterminated in both countries by intensive trapping programmes (Sheail 1988). The other, the mink (*Mustela vison*), is already widespread in the British Isles (Arnold 1984), and there has been speculation that it has had a severe impact, especially on waterside birds and mammals (eg Woodroffe, Lawton & Davidson 1990).

Ornamental species

The many species of ornamental waterfowl and pheasants which have been introduced into Britain are thoroughly reviewed by Lever (1977). The majority are scarcely established away from carefully managed estates. The species listed in Table 4 have maintained feral populations for many years, at least in a small area. Those marked with an asterisk may be spreading into the wider British countryside. Only the Canada goose (*Branta canadensis*), the red-legged partridge (*Alectoris rufa*), the pheasant and, in its specialised pinewood habitat, the capercaillie (*Tetrao urogallus*) have established self-maintaining populations which affect other wildlife significantly.

Another category of 'ornamental' species which have occasionally escaped or been released are cage-birds. The budgerigar (*Melopsittacus undulatus*) has bred outside the confines of free-flight aviaries at least three times on the British mainland, but seems unable to survive the more severe of

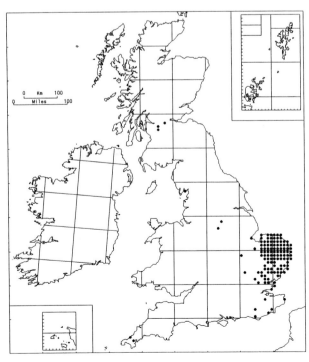

● 1955–1964

Figure 3. Records of coypu in Britain, 1955–64

British winters. A small population on the Scilly Isles may be more permanent (Lever 1977). The ring-necked parakeet (*Psittacula krameri*) is established in two or more parts of the London area, where flocks of 20 or more are frequently seen in autumn and winter.

Table 4. Introduced wildfowl and gamebirds in Britain. Species marked * are living feral and unaided; those marked ** are well established over large areas. Data from Lever (1977), Sharrock (1976) and Hollom (1975)

**	Canada goose (*Branta canadensis*)
	Egyptian goose (*Alopochen aegyptiaca*)
*	Mandarin duck (*Aix galericulata*)
	Wood duck (*Aix sponsa*)
*	Ruddy duck (*Oxyura jamaicensis*)
**	Capercaillie [reintroduction] (*Tetrao urogallus*)
**	Pheasant (*Phasianus colchicus*)
*	Golden pheasant (*Chrysolophus pictus*)
*	Lady Amherst's pheasant (*Chrysolophus amherstiae*)
*	Reeves's pheasant (*Syrmaticus reevesi*)
**	Red-legged partridge (*Alectoris rufa*)
	Chukar (*Alectoris chukar*)
	Bobwhite quail (*Colinus virginianus*)

The number of exotic plants introduced to Britain for horticulture runs into thousands, and many hundreds of species are permanently established in gardens. Of these, many occur as casuals on waste ground, roadsides and rubbish dumps, but few have invaded native plant communities. Among the exceptions are a group of three waterside plants which have successfully colonised many British rivers: giant hogweed (*Heracleum mantegazzianum*), Himalayan balsam (*Impatiens glandulifera*) and monkeyflower (*Mimulus guttatus*). One of the most successful ornamental alien plants, and certainly the one which has provoked the most active control measures from conservationists, is the rhododendron; details of its spread and habitat occupancy are given later.

Sporting

One of the few fish introductions which seems likely to have a widespread effect is that of the zander or pike-perch (*Stizostedion lucioperca*). This species has no close equivalent in Britain's impoverished native piscifauna. A sea barrier might be more effective against a freshwater fish than against almost any other organism. Unless a means of cross-Channel dispersal exists (it seems unlikely that fish eggs would stick to waterfowl feet and also be able to resist desiccation), all the species which were slow to spread in the post-glacial were unable to reach Britain.

The most carefully maintained and regularly restocked introduced fish is probably the rainbow trout (*Salmo gairdneri*). This species can affect the native fauna of a river considerably; but how much of its ecological impact (dragonflies and freshwater gastropod molluscs occur at very low densities or may be completely absent from stocked waters) is due to artificially high population levels rather than to intrinsic features of alien species is not known.

Less conspicuously alien, but perhaps of more ecological significance, are birds of prey lost or deliberately released by falconers. The present breeding population of goshawks (*Accipiter gentilis*) in Britain is thought to result largely or entirely from this source (Sharrock 1976); Kenward (1974) suggested that 50% of all goshawks kept by members of the British Falconers' Club were lost or released. Other species of raptor are doubtless 'topped up' from formerly captive birds. Whether this number is greater or less than those illegally removed from the wild by falconers and others is unknown.

In addition to the impacts of introduced gamebirds and fish, the effect of gamekeeping on wildlife is also extremely important: forage crops sown for pheasants also support flocks of native birds and small mammals, and so attract wintering raptors such as hen harriers (*Circus cyaneus*), merlins (*Falco columbarius*) and sparrowhawk (*Accipiter nisus*) (Marshall *et al.* 1989).

Other motives

The grass carp (*Ctenopharyngodon idella*) perhaps represents a non-introduction. It is apparently unable to breed freely in Britain but it is used to manage water-weed, so it has an impact on the countryside, in just the same manner as farm livestock or as sheep used to maintain short turf on nature reserves.

The motives behind the introduction of the little owl (*Athene noctua*) were very clearly stated by those who made the introduction. For instance, Waterton, who introduced the species to his estate in Yorkshire in 1842, thought that they would be 'particularly good for the horticulturalist in his kitchen gardens'. Lt-Col E G B Meade-Waldo, who introduced little owl into Kent between 1874–80, did so in order 'to rid belfries of sparrows and bats, and fields of mice'. Eight years later, Lord Lilford, in Northants, wrote in their favour: 'they are excellent mouse-catchers, very bad neighbours to young sparrows in their nests, and therefore valuable friends to farmers and gardeners'.

By the 1930s, the little owl was well established, and had been branded a 'menace' by the Press, and was being blamed for supposed declines in songbirds such as blackbirds (*Turdus merula*) and nightingales (*Luscinia megarhynchos*), and was alleged to take gamebird chicks. The controversy became so heated, with pro- and anti-little owl factions arguing emotively, but with scant evidence, that in 1935 the British Trust for Ornithology established the Little Owl Food Inquiry. In the report of the Inquiry, the Special Committee explained that 'wide currency . . . has been given to statements that the little owl is a wholesale destroyer of game-chicks, poultry-chicks and song birds' (Collinge *et al.* 1937). The analyst, Alice Hibbert-Ware, collated a large volume of varied correspondence, and dissected many hundreds of pellets, finally concluding that 'little owls feed almost wholly upon such insects, other invertebrates and small mammals as can readily be picked up on the ground during the hours of feeding – largely from dusk . . . to early morning' (Hibbert-Ware 1937, 1938).

The introduction of species for personal motives, or merely as a hobby, was not entirely the preserve of a small band of eccentrics. The 'Society for the Acclimatisation of Animals, Birds, Fishes, Insects and Vegetables within the United Kingdom' flourished briefly in the 1860s, but its French counterpart, 'Le Societé Impériale d'Acclimatation', was apparently much more prestigious and more widely accepted: in 1861, it boasted over 2000 members, including the Emperor Napoleon III and Pope Pius IX.

LIMITS TO THE SPREAD OF ALIENS

The difficulty in evaluating the success or otherwise of particular introductions, and the possible effects on native species, are twofold. First, naturalists often ignore aliens completely, unless specifically requested for information, and this disregard is exacerbated by the exclusion of alien species from most identification guides, and the fact that no-one is quite sure of their origin. Several aliens found in Britain have proved to be new to science, and at least three are still known only from Britain: the liverwort *Telaranea murphyae* is locally abundant on the Scilly Isles; the plant bug *Neodicyphus rhododendri* probably originates in North America (Dolling 1972; McGavin 1982); the snail *Gulella io* (Kerney & Cameron 1979) is probably from tropical Africa, but is known only from hothouses in Britain and Czechoslovakia. Several species of millipede (Blower 1985) and woodlouse (Harding & Sutton 1985) also have been described from hothouses, especially at Kew, but have not been collected in the wild anywhere in the world. Should we conserve them? They could be extinct in the wild! The other difficulty in evaluating the impact of an alien is that the 'evidence' is almost always anecdotal, and based on people's initial prejudices. Considering the little owl controversy of the 1930s, one could question what solid evidence there is for the supposed effects of mink on waterfowl and small mammal populations, though there is some evidence of changes in water vole (*Arvicola terrestris*) behaviour in the presence of mink (Woodroffe *et al.* 1990). Even well-known instances, such as introduced carnivores destroying native birds in New Zealand, may be open to less sensational interpretations (King 1984).

Climatic limits

Many alien species have established in the British Isles, but have not become widespread. In a few cases, the mechanism of this range restriction is known. Hottentot fig (*Carpobrotus edulis*), for instance, is limited by temperature, especially by frosts (Figure 5) (Preston & Sell 1988); in such a case, there is clearly potential for an expansion if the British climate warms.

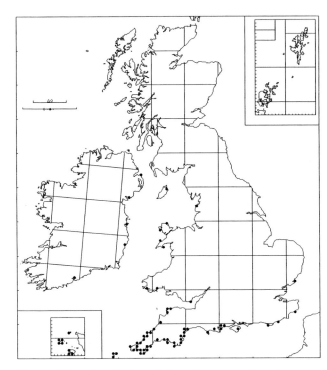

Figure 5. The distribution of *Carpobrotus edulis* in the British Isles (from Preston & Sell 1988)

Although the physiological cause has not been investigated, the Mediterranean snail (*Trochoidea elegans*) is probably likewise restricted; it is able to survive on south-facing chalk slopes, but rarely establishes elsewhere. Another snail, *Theba pisana*, has nearly all its recorded populations in or near car-parks in south-west England; this recording is not likely to be a mere reflection of conchologists' behaviour, so may imply some specific thermal characteristic of such dry, gravelly sites.

The only scorpion to become established in Britain, *Euscorpius flavicauda*, has at present a very restricted distribution, being confined to a handful of man-made sites, the best-known being Ongar railway station. Brickwork and concrete may have some resemblance to the dry, sun-baked soil or rock of the scorpion's native habitats in southern Europe. A small increase in summer temperatures could see the further spread of this remarkable exotic.

Habitat limits

On the edge of Britain's largest lowland raised mire, Thorne Moors in South Yorkshire, one of the country's first 'Garden Centres' was established in the early 1830s (Limbert 1991). By 1860, vast numbers of ericaceous shrubs were being grown: thousands of rhododendrons were raised from seed each year. A catalogue dated 1872 refers to 197 taxa of rhododendron, plus many hybrids, and 37 shrub genera. Three species survived this century: Labrador-tea (*Ledum palustre*) became extinct in about 1950; sheep-laurel (*Kalmia angustifolia*) survives and is dominant over an area of about 0.5 ha; rhododendron has become abundant over about 500 ha, nearly all of which is peripheral. Figure 6 shows this

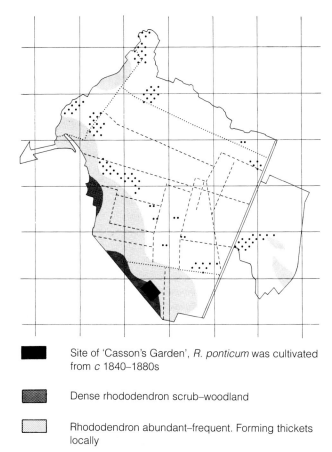

Site of 'Casson's Garden', *R. ponticum* was cultivated from *c* 1840–1880s

Dense rhododendron scrub–woodland

Rhododendron abundant–frequent. Forming thickets locally

Scattered bushes only

Figure 6. The spread of *Rhododendron ponticum* at Thorne Moors, South Yorkshire, Solid shading shows the location of 'Casson's Garden', a nursery where many *Rhododendron* species were cultivated *c* 1832–80; cross-hatching shows the area where *R. ponticum* is now the dominant shrub, to the exclusion of native species; simple hatching shows the area in which *R. ponticum* is frequent among native species; dots show isolated occurrences of single bushes of *R. ponticum* in native vegetation

area as a band along the western moor edge. The species is frequent, among native species, over about a third of the moors, with scattered individuals beyond. Rhododendron is almost absent from the central area (which was, until recent drainage, the wettest, and was covered with native vegetation). Drainage of the mire has been greatest near the edges, and perhaps the effects of fire (most severe in the drier areas) created an opening for rhododendron.

The spread of rhododendron in sand dunes, a very different habitat, has been documented by Fuller and Boorman (1977) using aerial photographs. They suggest that its establishment on the dunes was assisted by man's disturbance of the native vegetation, in this case through army activities in the 1940s creating open bare sand where seedlings could establish more easily.

Successful alien species might be expected to be generalists, able to exploit a wide range of conditions. However, some species are confined to a very narrow range of habitats. For instance, the New

Zealand willowherb occurs along the banks of streams and on wet rock faces in upland areas, and the very common ivy-leaved toadflax (*Cymbalaria muralis*) is almost confined to walls, though is occasionally found on shingle banks (Tutin *et al.* 1972; Philp 1982). It is unlikely that such aliens will compete with more than a handful of native species; and, in the case of *Cymbalaria*, there is no native species which occurs as frequently on the vertical faces of sunny walls.

Limits by dispersal

A few locally successful aliens appear to be limited by their inability to disperse. One unusual example is the tomato (*Lycopersicum esculentum*), which rarely produces seed in Britain, but is topped up from culinary sources via sewage works. Its distribution on river systems is thus almost always a downstream movement from seed sources.

The edible dormouse (*Glis glis*) is sufficiently well established in a small area of Hertfordshire to cause such domestic nuisance that Chiltern District Council has organised a trapping programme, and for the Forestry Commission to plan for its extermination as a pest (it strips bark from conifers); but, despite their fears, it seems not to have expanded its range since the 1940s (Lever 1977).

Regional variations

Even if a species is able to disperse and become established over a large part of the country, its frequency and impact may vary considerably from area to area. Three alien plants of wetlands, Himalayan balsam, giant hogweed and monkeyflower, are found beside lowland rivers and in waste places throughout much of Britain. In northern England, these species have become dominant over large stretches of industrial rivers (Figure 7; from Graham (1988)). It has been suggested that they are particularly successful where the native flora is suffering from the effects of pollution.

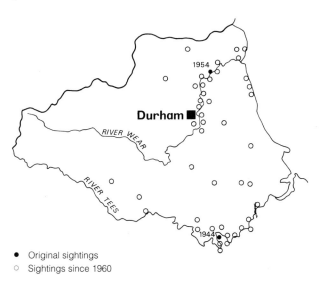

● Original sightings
○ Sightings since 1960

Figure 7. Spread of giant hogweed (*Heracleum mantegazzianum*) in County Durham (from Graham 1988)

One very specific form of freshwater pollution certainly assists aliens: thermal pollution. Warm-water outflows from power stations are able to support a number of species which cannot survive long term in cooler natural waters. These include the guppy or mosquito-fish (*Lebistes reticulatus*), and the submerged macrophyte *Vallisneria*. Two alien species of *Physa* (pond snails) are most abundant in artificially warmed water, but are able to survive in shallow ponds and ditches elsewhere.

The effects of warm water in rivers may occasionally extend into terrestrial habitats. Naturalised fig trees (*Ficus carica*) in the city of Sheffield are confined to the banks of the River Don, and to those parts which were formerly warmed by industrial coolant (Gilbert & Pearman 1988). All the recorded fig trees are mature, and no recent establishment has been noted. It is suggested that the trees established at a time when the Don flowed at a constant temperature of 20°C, and that, following the demise of the steel industry in the 1980s, temperatures are now too low for seedlings to establish, but mature trees are able to survive.

Ecologists' perception of the impact of an alien will be coloured by the habitat in which it is established. The rabbit, for instance, plays a vital role in maintaining open heathland on the Breck and short turf on calcareous grasslands, and these are scarce habitats valued by conservationists. Its activity in woodlands, moorlands, wetter grassland and in agricultural areas is far less sympathetically received.

Alien species, as much as natives, tend to occupy different habitats in different parts of their range. Many invertebrate species become coastal in the northern part of their range, as the low-lying coastal strip in eastern Scotland presumably has a milder climate. This fact applies as strongly to introduced molluscs, such as *Helix aspersa*, as to native ones. A wide range of molluscs which are garden species in central and south-east England are confined to glasshouses in Scotland, but are free-living in 'wild' habitats such as woodland in south-west England and in Ireland (Kerney & Cameron 1979; Eversham & Baxter 1989).

Several synanthropic invertebrates are able to survive out-of-doors in southern England. House spiders (*Tegenaria* spp.) occur in disused quarries, the house cricket (*Acheta domesticus*) may survive in refuse heaps, and the silverfish (*Lepisma saccharina*) occurs on rocks and walls, browsing on lichens and algae.

These unusual occurrences may be a simple effect of temperature, in which case global warming may cause a marked expansion of the habitats occupied by such aliens.

IMPACT OF ALIENS ON NATIVE SPECIES AND HABITATS

If a species has a very close native ally, either congeneric or with a similar ecological role, there is more chance of direct competition, and greater potential for the native to suffer if the alien flourishes.

Theories of co-evolution should favour the native species, but the absence of biotic checks on the alien could tip the balance in the other direction: unless the alien is phylogenetically very close to the native species, it may not share its predators or, especially, parasites and parasitoids.

There are surprisingly few documented cases of direct competition between a native and an alien species. The moss *Orthodontium lineare*, a 20th century colonist from the southern hemisphere, may be ousting *O. gracile*, as they occupy a similar microhabitat, although the alien is much more eurytopic (Smith 1977; Rose & Wallace 1974).

The most famous case of possible competition between native and alien species is perhaps that of the red and grey squirrel (*Sciurus vulgaris* and *S. carolinensis*). Kenward and Tonkin (1986) suggested that the grey is better able to digest acorns and beech mast, so is at a competitive advantage in deciduous woodland, whereas the native red is better adapted to conifers. However, if the evidence in the case of well-studied animals such as the squirrel is inconclusive, it would be correspondingly more difficult to prove competition in other, less conspicuous, species.

The question has often been asked: are some habitats much more invasible than others? And, if so, why?

Crawley (1987) tabulated the habitats of native and alien plant species, as given in the standard *Flora* (Clapham, Tutin & Warburg 1962), reproduced as Table 5. This Table shows that man-made habitats have far more aliens than semi-natural habitats. The same applies to molluscs (Table 6, derived from the modern *Field guide* (Kerney & Cameron 1979)), although there are a few differences between habitats. That these differences are not just artifacts caused by personal biases of flora and mollusc field guide writers can be shown by examining the total flora and mollusc fauna of well-recorded individual sites.

Table 5. The proportion of the vascular flora of selected habitats in Britain which are aliens. Data from Crawley (1987)

Class	Habitat	% alien
Man-made	Wasteland	78
	Walls	46
	Fields	37
	Hedgerow	22
Woodland	Conifer plantation	56
	Deciduous	5
Wetland	Bog	5
	Fen	2
Grassland	Sea cliffs	18
	Damp grassland	13
	Dunes	13
	Dry grass	5

Table 6. The proportion of alien and native land molluscs in selected habitats. Data from Kerney and Cameron (1979)

Class	Habitat	No. native	No. alien	% alien
Man-made	Glasshouses	4	15	79
	Agricultural	5	10	67
	Garden	13	18	58
	Wasteland	5	6	55
	Hedgerow	25	8	29
	Walls	10	3	23
Woodland	Deciduous	63	9	13
	Scrub	7	1	13
	Conifer plantation	7	0	0
Wetland	Marsh or fen	29	0	0
	Bog, moorland	3	0	0
Grassland	Dunes	14	7	33
	Dry grassland	15	5	25
	Damp grassland	18	3	14

Table 7. The native and alien flora of a range of sites. Data from Steele and Welch (1973), Rixon and Peterken (1975), Sage (1966), George (1961), Brookes and Burns (1970), Harding *et al.* 1988, and Payne (1978)

Site	Native	%	Alien	%
Monks Wood NNR	349	94.3	21	5.7
Bedford Purlieus SSSI	439	93.8	29	6.2
Northaw Great Wood	241	90.3	26	9.7
Dale Parish	434	88.9	54	11.1
Slapton Ley NNR	429	86.8	65	13.2
Barking Reach	162	74.3	56	25.7
Essex walls: total	203	71.0	83	29.0
on 5% of walls	31	66.0	16	34.0

Table 8. The native and alien mollusc fauna of a range of sites. Data from Steele and Welch (1973), Cameron (1978), Stratton (1964), Eversham (1991) and unpublished lists in the possession of BCE

Site	Native	%	Alien	%
Monks Wood NNR	35	94.5	4	5.4
Malham area	59	93.7	4	6.3
UEA Fen	25	92.6	2	7.4
Dale parish	30	88.2	4	11.8
Lindholme	30	85.7	5	14.3
Huntingdon (garden)	13	56.5	10	43.5

Tables 7 and 8 derive from a series of sites for which comprehensive plant and mollusc lists have been published. They show precisely the same pattern as the generalised lists. One difference is apparent: there are far fewer alien plants at some sites than others, whereas the numbers of species of molluscs is fairly similar at all but very synanthropic sites. The proportion of aliens in the lists obviously differs depending on the size of the native fauna. For example, aliens are much more important in disturbed habitats than in ancient native woodland.

WHAT MAKES A SUCCESSFUL INVADER?

Lawton and Brown (1987) correlated body size with the proportion of successful invasions in a taxonomic group. They pointed out that the biology of large species tends to be better understood, and so people are less tempted to try the impossible. At the same time, they also emphasised population parameters (r – intrinsic rate of population increase, and K – carrying capacity) as a possible explanation of the relative success in establishing.

A more direct measure of the attention paid to a group of organisms is the amount published on the group each year. A readily available estimate is a count of the number of pages of the *Zoological Record*. If the number of pages is divided by the number of species in the group, it correlates well with the success of the group in establishing (Figure 8). So, at the higher taxonomic levels, better-known organisms are more likely to be successful when introductions are attempted, presumably because very unlikely attempts would be discounted in advance of the attempt.

Lawton and Brown (1987) noted that insects show the reverse trend – orders with mostly small species do best. An alternative explanation of this effect relies on the global distribution of the major insect orders (Figure 9). The orders with a high proportion of species living in the north temperate zone are the most successful colonists of Britain. The only large group omitted from Figure 9 are the Hymenoptera: a high proportion of described species are obligate, narrowly specific parasitoids, which cannot be expected to behave similarly to most insects. Not surprisingly, Hymenoptera lie far below the regression line of Figure 9.

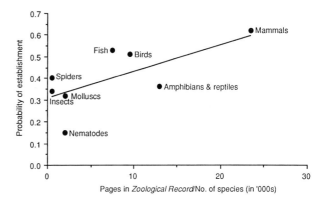

Figure 8. The relationship between the likelihood of successful establishment of alien species in Britain, and the volume of zoological literature devoted to the group (expressed as pages in *Zoological Record* in 1988, divided by the number of species in the taxonomic group)

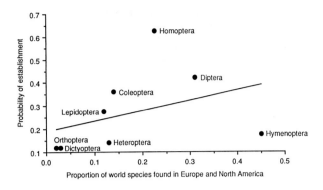

Figure 9. The relationship between the likelihood of successful establishment of alien species in Britain, and the proportion of the world fauna in the taxonomic group which is native to Europe and North America

The fact that all the estimates of world totals are gross underestimates is not a problem. As with phylum differences, a crucial factor is how well watched a group is: it is uncertain whether anyone would notice if an alien symphylan was rampaging across southern England, for instance. The most marked failures to establish are in the most thermophilous groups – amphibians, reptiles, Dictyoptera. In the latter, the documented casuals are all large tropical species; there is no recorded occurrence of an alien *Ectobius* cockroach.

Rather than seek general principles in terms of population parameters, the pattern of success and failure may be better described in terms of origins and habitats.

Crawley's list (Table 9) usefully discounted life history strategy as the key to alien success: the 'top 20' aliens include annuals (eg Canadian fleabane

(*Erigeron canadensis*)) biennials (beaked hawk's-beard (*Crepis vesicaria*)), and long- and short-lived perennials (such as American willowherb (*Epilobium ciliatum*) or sycamore (*Acer pseudoplatanus*)). Table 9 also demonstrates the diversity of dispersal mechanisms in highly successful aliens: wind-dispersed Compositae, sycamore and willowherbs, explosive pods in Himalayan balsam and rhododendron, succulent berries in snowberry (*Symphoricarpos albus*), and heavy smooth seeds in the umbellifers, whose means of dispersal are unclear. Although the list may be a little subjective, it certainly includes most of the very successful invaders of natural communities – the majority of nature reserves in Britain now seem to contain snowberry and ground elder (*Aegopodium*).

SURPRISING FAILURES AND EVENTUAL SUCCESSES

The case of the edible dormouse has already been mentioned: it is well established, and has expanded a little. It is unclear why it is not able to spread further.

The local success of porcupines (*Hystrix* sp.) in south-west England (Lever 1977; Smallshire & Davey 1989) poses the same question: if an animal is able to reach pest status in one area, why does it not become more widespread? The answer in this case, as in that of the edible dormouse, may be its lack of dispersive power in the English landscape.

These aliens are completely at home in the British climate, and are known to be able to breed successfully, and to live long adult lives. Ferrets (*Mustela furo*) can do this, and have the additional advantage of high mobility. Yet there is little factual evidence of

Table 9. The 'top 20' British alien plants (from Crawley 1987)

Species	Family	Habitat
Acer pseudoplatanus	Aceraceae	Woodland
Aegopodium podagraria	Umbelliferae	Gardens, wasteland, etc
Avena fatua	Graminae	Cultivated land
Buddleia davidii	Buddleiaceae	Wasteland, railways, walls
Centranthus ruber	Valerianaceae	Walls, cliffs
Crepis vesicaria	Compositae	Roadsides, wasteland
Elodea canadensis	Hydrillidae	Slow water
Epilobium brunnescens	Onagraceae	Streamsides
E. ciliatum	Onagraceae	Gardens, wasteland, etc
Erigeron canadensis	Compositae	Cultivated, railways, etc
Impatiens glandulifera	Balsaminaceae	Riversides, fens, carr
Matricaria suaveolens	Compositae	Tracks, etc
Mimulus guttatus	Scrophulariaceae	Rivers and streams
Reynoutria japonica	Polygonaceae	Roadsides, wasteland
Rhododendron ponticum	Ericaceae	Woodland, heathland
Senecio squalidus	Compositae	Wasteland, railways, walls
Smyrnium olusatrum	Umbelliferae	Roadsides and cliffs by the sea
Symphoricarpos albus	Caprifoliaceae	Shady wasteland, woods
Veronica filiformis	Scrophulariaceae	Lawns, etc
V. persica	Scrophulariaceae	Cultivated land

established feral populations, other than on a small number of offshore islands (Mull, Isle of Man).

The dice snake (*Natrix tesselatus*) has been predicted by several ecologists as a potential colonist, being ecologically like a grass snake (*Natrix natrix*) but rather more aquatic, and often kept by amateur herpetologists in captivity. The chances of escape thus provide a means of introduction.

The fact that the first two attempts to introduce little owls to Britain failed shows clearly that a single failure cannot be taken to indicate that a species is unsuitable or unable to compete with the resident biota. The role of chance in establishment should not be underestimated.

WHAT OF THE FUTURE?

There are more man-made habitats in Britain than ever before, providing greater scope for invasion. Likewise, the increase in disturbance to natural and semi-natural vegetation, caused by an increasingly mobile and affluent human population, and the gradual loss of indigenous species through land use change, pollution, pesticides, etc, may all make the British countryside more accessible for alien species to become established.

At the risk of jumping on an already over-crowded bandwagon, global warming will remove climatic constraints on some species, and probably increase their dispersal. Its effect on native populations and plant communities may well also increase habitat penetration by aliens.

Thus, there is a great need for more careful surveillance; BRC and the national recording schemes provide the means to carry out surveillance and to analyse the results.

REFERENCES

Anon. 1986. Reintroductions (FFPS Symposium, 1 December 1984). *Mammal Review,* **16(2)**, 49–88

Akeroyd, J.R., Warwick, S.I & Briggs, D. 1978. Variation in four populations of *Senecio viscosus* L. as revealed by cultivation experiments. *New Phytologist,* **81**, 391–400.

Arnold, E.A. & Burton, J.A. 1978. *Field guide to the reptiles and amphibians of Britain and Europe.* London: Collins.

Arnold, H.R. 1984. *Distribution maps of the mammals of the British Isles.* Huntingdon: Biological Records Centre. (Unpublished.)

Baker, G.H. 1989. Damage, population dynamics, movement and control of pest helicid snails in southern Australia. In: *Slugs and snails in world agriculture,* edited by I.F. Henderson, 175–186. (*British Crop Protection Council Monograph, 41.*) Bracknell: BCPC Publications.

Barker, G.M. 1989. Slug problems in New Zealand pastoral agriculture. In: *Slugs and snails in world agriculture,* edited by I.F. Henderson, 59–58. (*British Crop Protection Council Monograph, 41.*) Bracknell: BCPC publications.

Barrett, B.I.P., Byers, R.A. & Bierlein, D.L. 1989. Conservation tillage crop establishment in relation to density of the field slug (*Deroceras reticulatum* (Muller)). In: *Slugs and snails in world agriculture,* edited by I.F. Henderson, 93–100. (*British Crop Protection Council Monograph, 41.*) Bracknell: BCPC publications.

Barrett-Hamilton, G.E.H. & Hinton, M.A.C. 1910–21. *A history of British mammals.* London: Gurney & Jackson.

Barrington, R.M. 1880. On the introduction of the squirrel into Ireland. *Scientific Proceedings of the Royal Dublin Society,* N.S. **2**, 615–631.

Berry, R.J. 1963. Epigenetic polymorphisms in wild populations of *Mus musculus. Genetical Research,* **4**, 193–220.

Berry, R.J. 1968. The ecology of an island population of the house mouse. *Journal of Animal Ecology,* **37**, 445–470.

Berry, R.J. 1970. Covert and overt variation, as exemplified by British mouse populations. *Symposium of the Zoological Society of London,* **26**, 3–26.

Berry, R.J. & Jakobson, M.E. 1974. Vagility in an island population of the house mouse. *Journal of Zoology, London,* **173**, 341–354.

Blower, J.G. 1985. *Millipedes.* Synopses of the British fauna (New Series), no. 35. London: Brill/Backhuys.

Bolton, B. & Collingwood, C.A. 1975. Hymenoptera: Formicidae. (*Handbooks for the identification of British insects, Vol. VI, Part 3(c).*) London: Royal Entomological Society.

Boorman, L.A., & Fuller, R.M. 1981. The changing status of reedswamp in the Norfolk Broads. *Journal of Applied Ecology,* **18**, 241–269.

Bretherton, R.F. 1989. The continental swallowtail, *Papilio machaon gorganus.* In: *The moths and butterflies of Great Britain and Ireland: Vol. 7, Part 1, Hesperiidae – Nymphalidae,* edited by A.M. Emmet & J. Heath, 78–80. Colchester: Harley Books.

Brookes, B.S. & Burns, A. 1970. The natural history of Slapton Ley nature reserve: III. The flowering plants and ferns. *Field Studies,* **3**, 125–157.

Buckland, P.C. 1979. *Thorne Moors: a palaeoecological study of a Bronze Age site.* (Occasional publication no. 8.) Birmingham: University of Birmingham, Department of Geography.

Buckland, P.C. & Johnson, C. 1983. *Curimopsis nigrita* (Palm) [Coleoptera: Byrrhidae] from Thorne Moors, South Yorkshire. *Naturalist,* **108**, 153–154.

Cameron, R.A.D. 1978. Terrestrial snail faunas of the Malham area. *Field Studies,* **4**, 715–728.

Clapham A.R., Tutin, T.G. & Warburg, E.F. 1962. *Flora of the British Isles.* 2nd ed. Cambridge: Cambridge University Press.

Clapham, A.R., Tutin, T.G. & Moore, D.M. 1987. *Flora of the British Isles.* 3rd ed. Cambridge: Cambridge University Press.

Collinge W.E., Fryer, J.C.F., Jourdain, F.C.R. & Kinnear, N.B. 1937. Report of the Special Committee [of the Little Owl Food Inquiry]. *British Birds,* **31**, 162–163.

Collins, N.M. & Wells, S.M. 1987. *Invertebrates in need of special protection in Europe.* (Council of Europe nature & environment series no. 35.) Strasbourg: Council of Europe.

Corbet, G. & Ovenden, D. 1980. *The mammals of Britain and Europe.* London: Collins.

Cramp, S. 1977. *Handbook of the birds of Europe, the Middle East and North Africa, the birds of the Western Palaearctic, Vol. 1.* Oxford: Oxford University Press.

Cramp, S. 1980. *Handbook of the birds of Europe, the Middle East and North Africa, the birds of the Western Palaearctic, Vol. 2.* Oxford: Oxford University Press.

Cramp, S. 1983. *Handbook of the birds of Europe, the Middle East and North Africa, the birds of the Western Palaearctic, Vol. 3.* Oxford: Oxford University Press.

Cramp, S. 1985. *Handbook of the birds of Europe, the Middle East and North Africa, the birds of the Western Palaearctic, Vol. 4.* Oxford: Oxford University Press/Royal Society for the Protection of Birds.

Cramp, S. 1988. *Handbook of the birds of Europe, the Middle East and North Africa, the birds of the Western Palaearctic, Vol. 5.* Oxford: Oxford University Press/Royal Society for the Protection of Birds.

Crawley, M.J. 1987. What makes a community invasible? In: *Colonisation, succession and stability,* edited by A.J. Gray, M.J. Crawley & P.J. Edwards, 429–453. Oxford: Blackwell Scientific.

Cross, J.R. 1975. *Rhododendron ponticum* L. (Biological flora of the British Isles, no. 137.) *Journal of Ecology,* **63**, 345–364.

Crossley, R. & Norris, A. 1975. *Bembidion humerale* Sturm (Col., Carabidae) new to Britain. *Entomologist's Monthly Magazine,* **111**, 59–60.

Dolling, W.R. 1972. A new species of *Dicyphus* from southern England. *Entomologist's Monthly Magazine,* **107**, 244–245.

Elton, C.S. 1958. *The ecology of invasions by animals and plants.* London: Methuen.

Enghoff, H. 1982. The millipede genus *Cylindroiulus* on Madeira – an insular species swarm (Diplopoda, Julida, Julidae). *Entomologica Scandinavica, Supplement,* **18**, 1–142.

Eversham, B.C. 1991. Changes in the mollusc fauna of the Huntingdon garden. *Report of the Huntingdon Fauna and Flora Society,* 43rd, 1990, 32–38.

Eversham, B.C & Baxter J. 1989. Mollusca. In: *Survey of specialist biological groups in Northern Ireland,* edited by P.T. Harding. (NERC contract report to the Department of the Environment for Northern Ireland.) Huntingdon: Institute of Terrestrial Ecology.

Fisher, J. 1966. The fulmar population in Britain and Ireland, 1959. *Bird Study,* **13**, 5–76.

Foltz, D.W., Ochman, H., Jones, J.S., Evangelisti, S.M. & Selander, R.K. 1982 Genetic population structure and breeding systems in arionid slugs (Mollusca: Pulmonata). *Biological Journal of the Linnean Society,* **17**, 225–242.

Fuller, R.M. & Boorman, L.A. 1977. The spread and development of *Rhododendron ponticum* L. on dunes at Winterton, Norfolk, in comparison with invasion by *Hippophae rhamnoides* L. at Saltfleetby, Lincolnshire. *Biological Conservation,* **12**, 83–94.

George, M. 1961. The flowering plants and ferns of Dale, Pembrokeshire. *Field Studies,* **1**, 1–24.

Gilbert, O.L. & Pearman, M.C. 1988. Wild figs by the Don. *Sorby Record,* **25**, 31–33.

Godwin, H. 1975. *The history of the British vegetation.* Cambridge: Cambridge University Press.

Gosling, L.M. & Baker, S.J. 1989. The eradication of muskrats and coypus from Britain. *Biological Journal of the Linnean Society,* **38**, 39–51.

Graham, G.G. 1988. *The flora and vegetation of County Durham.* Durham: Durham Flora Committee/Durham County Conservation Trust.

Gray, A.J., Crawley, M.J. & Edwards, P.J., eds. 1987. *Colonization, succession and stability.* (British Ecological Society Symposium no. 26.) Oxford: Blackwell Scientific.

Hall, M.L. & Emmet, A.M. 1989. The swallowtail. *Papilio machaon britannicus.* In: *The moths and butterflies of Great Britain and Ireland: Vol. 7 Part 1, Hesperiidae – Nymphalidae,* edited by A.M. Emmet & J. Heath, 76–78. Colchester: Harley Books.

Harding, P.T. 1978. *A bibliography of the occurrence of certain woodland Coleoptera in Britain with special reference to timber-utilising species associated with old trees in pasture-woodlands.* (CST report no. 161.) Banbury: Nature Conservancy Council.

Harding, P.T., Eversham, B.C., Preston, C.D. & Hooper, M.D. 1988. *Terrestrial ecological studies at Barking Reach.* (NERC contract report to Thames Power Ltd.) Huntingdon: Institute of Terrestrial Ecology.

Harding, P.T. & Sutton, S.L. 1985. *Woodlice in Britain & Ireland: distribution and habitat.* Huntingdon: Institute of Terrestrial Ecology.

Harvie-Brown, J.A. 1880–81. The squirrel in Great Britain. *Proceedings of the Royal Physical Society of Edinburgh,* **5**, 343–348; **6**, 31–63, 115–182.

Heath, J., Pollard, E. & Thomas, J.A. 1984. *Atlas of butterflies in Britain and Ireland.* Harmondsworth: Viking.

Hengeveld, R. 1989. *Dynamics of biological invasions.* London: Chapman & Hall.

Hibbert-Ware, A. 1937. Report of the Little Owl Food Inquiry, 1936–37. *British Birds,* **31**, 162–187.

Hibbert-Ware, A. 1938. Report of the Little Owl Food Inquiry, 1936–37. *British Birds,* **31**, 205–229, 249–264.

Higgins, L.G. & Riley, N.D. 1983. *A field guide to the butterflies of Britain and Europe.* London: Collins.

Hollom, P.A.D. 1975. *Popular handbook of British birds.* London: Witherby.

Huntley, B. & Birks, H.J.B. 1983. *An atlas of past and present pollen maps of Europe, 0–13000 years ago.* Cambridge: Cambridge University Press.

Jalas, J. & Suominen, J. 1972–86. *Atlas Florae Europaeae, Vols 1–7.* Helsinki: Committee for Mapping the Flora of Europe and Societas Biologica Fennica Vanamo.

Jessen, K. & Halbaek, H. 1944. Cereals in Great Britain and Ireland in prehistoric and early historic times. *Kongelige Danske Videnskabernes Selskabs,* **3** (2), 1.

Johnson, C. 1978. Notes on Byrrhidae (Col.); with special reference to, and a species new to, the British fauna. *Entomologist's Record and Journal of Variation,* **90**, 141–147.

Kenward, R.E. 1974, Mortality and fate of trained birds of prey. *Journal of Wildlife Management,* **38**, 751–756.

Kenward, R.E. & Tonkin, M. 1986. Red and grey squirrels; some behavioural and biometric differences. *Journal of Zoology,* (A), **209**, 279–281

Kerney, M.P. & Cameron, R.A.D. 1979. *A field guide to the land snails of Britain and north-west Europe.* London: Collins.

King, C. 1984. *Immigrant killers: introduced predators and the conservation of birds in New Zealand.* Auckland & Oxford: Oxford University Press.

Kornberg, H. & Williamson, M.H. 1987. *Quantitative aspects of the ecology of biological invasions.* London: Royal Society.

Lawton, J.H. & Brown, K.C. 1987. The population and community ecology of invading insects. In: *Quantitative aspects of the ecology of biological invasions,* edited by H. Kornberg & M.H. Williamson, 105–113. London: Royal Society.

Limbert, M. 1991. William Casson of Thorne. *Naturalist,* **116**, 3–15.

Lever, C. 1977. *The naturalized animals of the British Isles.* St Albans: Granada.

Luff, M.L. 1982. *Preliminary atlas of the British Carabidae (Coleoptera).* Huntingdon: Biological Records Centre.

Mabberley, D.J. 1987. *The plant-book, a portable dictionary of the higher plants.* Cambridge: Cambridge University Press.

Maitland, P.S. 1977. *The freshwater fishes of Britain and Europe.* London: Hamlyn.

Marshall, R.A., Lynes, M. & Limbert, M. 1989. *The vertebrate fauna of Hatfield Moors.* (Lapwing Special Series, 5.) Doncaster: Doncaster and District Ornithological Society.

May, R. & Fisher, J. 1953. A collared turtle dove in England. *British Birds,* **46**, 51–55.

McGavin, G.C. 1982. A new genus of Miridae (Hem.: Heteroptera). *Entomologist's Monthly Magazine,* **118**, 79–86.

Merritt, R., Moore, N.W. & Eversham, B.C. 1992. *Atlas of the dragonflies of Britain and Ireland.* London: HMSO. In press.

Mollison, D. 1986. Modelling biological invasions: chance, explanation, prediction. In: *Quantitative aspects of the ecology of biological invasions,* edited by H. Kornberg & M.H. Williamson, 173–189. London: Royal Society.

Morrill, J. 1988. The Stuarts. In: *The Oxford history of Britain,* edited by K.O. Morgan, 327–398. Oxford: Oxford University Press.

Morris, P. 1983. *Hedgehogs.* Surrey: Whittet.

Parslow, J.L.F. 1973. *Breeding birds of Britain and Ireland.* Berkhamsted: Poyser.

Paton, J.A. 1965. *Lophocolea semiteres* (Lehm.) Mitt. and *Telaranea murphyae* sp.nov. established on Tresco. *Transactions of the British Bryological Society*, **4**, 775–779.

Payne, R.M. 1978. The flora of walls in south-east Essex. *Watsonia*, **12**, 41–46.

Peterson, R.T., Mountford, G. & Hollom, P.A.D. 1983. *A field guide to the birds of Britain and Europe*. London: Collins.

Philp, E.G. 1982. *Altas of the Kent flora*. West Malling: Kent Field Club.

Pigott, C.D. 1981. The status and ecology of *Tilia platyphyllos* in Britain. In: *The biological aspects of rare plant conservation*, edited by H. Synge, 305–317. Chichester: Wiley.

Polunin, O. 1980. *Flowers of Greece and the Balkans: a field guide*. Oxford: Oxford University Press.

Preston, C.D. & Sell, P.D. 1988. The Aizoaceae naturalized in the British Isles. *Watsonia*, **17**, 217–245.

Rixon, P. & Peterken, G.F. 1975. Vascular flora. In: *Bedford Purlieus, its history, ecology and management*, edited by G.F. Peterken & R.C. Welch, 101–108. (Monks Wood Symposium no. 7.) Huntingdon: Institute of Terrestrial Ecology.

Rose, F. & Wallace, E.C. 1974. Changes in the bryophyte flora of Britain. In: *The changing flora and fauna of Britain*, edited by D.L. Hawksworth, 27–46. (Systematics Association special volume no. 6.) London: Academic Press.

Sage, B.L., ed. 1966. *Northaw Great Wood*. Welwyn: Hertfordshire County Council.

Sharrock, J.T.R. 1976. *The atlas of breeding birds in Britain and Ireland*. Calton: Poyser.

Sheail, J. 1971. *Rabbits and their history*. Newton Abbot: David & Charles.

Sheail, J. 1988. The extermination of the muskrat (*Ondatra zibethicus*) in inter-war Britain. *Achives of Natural History*, **15**, 155–170.

Skidmore, P., Limbert, M. & Eversham, B.C. 1987. The insects of Thorne Moors. *Sorby Record*, **23** (Supplement), 89–153.

Smallshire, D. & Davey, J.W. 1989. Feral Himalayan porcupines in Devon. *Nature in Devon*, **10**, 62–69.

Smith, A.J.E. 1977. Distribution maps of bryophytes: *Orthodontium gracile*. *Journal of Bryology*, **9**, 401.

Stebbings, R.E. 1988. *Conservation of European bats*. London: Christopher Helm.

Steele, R.C. & Welch, R.C., eds. 1973. *Monks Wood: a nature reserve record*. Huntingdon: Nature Conservancy.

Stratton, L.W. 1964. The non-marine Mollusca of the parish of Dale. *Field Studies*, **2**, 41–52.

Sutton, S.L. & Harding, P.T. 1989. Interpretation of the distribution of terrestrial Isopods in the British Isles. *Monitore zoologico italiano, (N.S.) Monografia*, **4**, 43–61.

Trevelyan, G.M. 1926. *History of England*. London: Longman.

Turrill, W.B. 1929. *The plant-life of the Balkan peninsula; a phytogeographic study*. Oxford: Oxford University Press.

Tutin, T.G., Heywood, V.H., Burges, N.A., Moore, D.M., Valentine, D.H., Walters, S.S. & Webb, D.A. 1972. *Flora Europaea, Vol. 3*. Cambridge: Cambridge University Press.

Twigg, G. 1975. *The brown rat*. Newton Abbot: David & Charles.

Webb, D.A. 1985. What are the criteria for presuming native status? *Watsonia*, **15**, 231–236.

Welch, R.C. 1987. Species account for *Limoniscus violaceus*. In: *British Red Data Books, 2: Insects*, edited by D.B. Shirt, 190–191. Peterborough: Nature Conservancy Council.

Wells, T.C.E., Bell, S.A. & Frost, A. 1981. *Creating attractive grasslands using native plants*. Shrewsbury: Nature Conservancy Council.

Woodroffe, G.L., Lawton, J.H. & Davidson, W.L. 1990. The impact of feral mink *Mustela vison* on water voles *Arvicola terrestris* in the North Yorkshire Moors National Park. *Biological Conservation*, **51**, 46–62.

Monitoring populations of a butterfly during a period of range expansion

E Pollard

Springhill Farm, Benenden, Cranbrook, Kent TN17 4LA

INTRODUCTION

There is little doubt that the distribution of all of our butterflies is in more or less continual flux, and that this has been true since they recolonised Britain after the last ice age, some 10 000 years ago. Any apparent stability is probably due to the short period over which observations have been made; only in the mid-19th century was the first attempt made to describe the approximate distribution of British butterflies (Fust 1868). Reasonably complete maps, produced by the Biological Records Centre, were not available until 1984 (Heath, Pollard & Thomas 1984).

There is also little doubt that post-glacial changes in distribution have been driven by two main factors, climate and man. The impact of man, through changes in and destruction of biotopes, has probably been of much greater importance than climate in recent years, but the effects of climatic shifts are certainly continuing. If predictions about global warming are correct, climatic change will shortly have a further major effect on our butterflies.

Previous studies on changes in range and changes in abundance have been speculative because, inevitably, the reliability of the information available has been poor. For a few species, the approximate timing of changes in range over the last century can be reconstructed from reports in entomological journals. Examples include the comma (*Polygonia c-album*) (Pratt 1986–87) and the white admiral (*Ladoga camilla*) (Pollard 1979a).

Similarly, some indication of changes in abundance can be obtained from published accounts which describe particular years as good or bad for butterflies; for example, Beirne (1955) made a comprehensive study of such information on abundance and related perceived changes in abundance to weather, although, clearly, statistical analysis was not possible.

Quantitative, synoptic information on fluctuations in the abundance of butterflies has been available only in very recent years. The butterfly monitoring scheme, supported by the Nature Conservancy Council and the Institute of Terrestrial Ecology, began in 1976 (Pollard 1979b). The scheme is based on simple counts along fixed routes at sites throughout Britain, and provides data on local distribution, on relative abundance and on phenology. Fluctuations in abundance are often associated with weather (Pollard 1988), but changes in the biotopes in which the butterflies live are also important (Pollard, Hall & Bibby 1986).

In a period when the ranges of many butterflies have contracted sharply (Heath *et al.* 1984), a few species have expanded their ranges. The latter, such as the speckled wood (*Pararge aegeria*), small skipper (*Thymelicus sylvestris*), and hedge brown (*Pyronia tithonus*), are common within their ranges, and so more comprehensive data are available than for rare or locally distributed declining species.

The hedge brown will be used as a case study in this account to show the nature of the information that is now available from distribution and monitoring schemes; it will also show how such schemes can complement each other.

The hedge brown is a univoltine satyrid butterfly which flies mainly in July and August. The larvae feed on a range of grass species and overwintering is in the larval stage. Its general natural history is described by Brakefield and Emmet (1989). It is a common butterfly in a wide range of biotopes, being most abundant where grassland and scrub occur together.

It seems reasonable to expect some association between abundance and distribution; for example, absolute abundance may be greater at the centre than at the edge of the range. Similarly, expansion of range might be associated with increase in abundance, and such increase dependent on weather which favoured the particular species. Indeed, this study began as an examination of possible effects of weather on abundance and range, although the emphasis changed during its progress, for reasons that will become clear.

RANGE

During the last 100 years or so, the range of the hedge brown in Britain has contracted southwards and subsequently re-expanded. Heath *et al.* (1984) noted these changes in outline, but Eversham (pers. comm.) has made more detailed analysis, using both the data on which the butterfly distribution maps are based and additional, more recent, information. The timing of the contraction of range is uncertain, but it appears to have reached its minimum range in the 1930s or 1940s in both Britain and Ireland. However, outlying populations persisted as the main boundary of the range retreated south. Some of these outliers survived until re-absorbed within the main range as the species re-expanded. At the minimum, the main northern boundary of the hedge brown was in the southern midlands of England, while in Ireland it was largely confined to southern coastal areas.

Subsequently, in both countries, the hedge brown has moved north again to occupy much, but not all, of the area it occupied in the last century. Again, the timing of the expansion is uncertain. To a large extent such uncertainty is inevitable; a new colony may be first observed only several years after the arrival of the first individual or individuals. The expansion appears to be continuing, and Eversham (pers. comm.) estimates that the main range edge may have moved about 50 km north since 1970. These additional analyses conducted by Eversham indicate that there is still much potential for exploring range changes using the BRC database. The published maps provide no more than a summary of these data.

ABUNDANCE

During the latter part of the recent expansion of range of the hedge brown, populations over much of its range in Britain have been monitored in the butterfly monitoring scheme. The butterfly has colonised only one site in the scheme, in the Derbyshire Peak District, and the first records there were in 1989. Thus the information obtained relates, effectively, only to sites within its range before the period of expansion and covers the period from 1976 to 1989.

The hedge brown has been recorded at virtually all sites in the monitoring scheme that are within its ranges. The data presented here are restricted to 35 sites where the species occurs in, at least, moderate numbers (except in years of general scarcity), and where index values are available for at least seven years over a period of at least ten years. For explanation of the method by which index values are derived, see Pollard *et al.* (1986).

Absolute abundance

The recording method used in the monitoring scheme is not designed to provide information on differences in abundance (total population size) between sites. The size of the counts at a site depends partly on the suitability for the species of the particular route chosen, partly on the length of the route, and partly on differences between recorders; thus, only a weak correlation with absolute abundance at a site can be expected even when differences in the lengths of route are taken into account. Nevertheless, a strong association between abundance and proximity to the edge of the range would be expected to be revealed.

The relationships between index values per unit length of transect route and (i) latitude, and (ii) distance from the edge of the range were examined for each of the 14 years from 1976 to 1989. The two measures of location are not equivalent, as the edge of the range extends much further north at the coasts than inland. However, they produce similar results and these are quite surprising. In 13 of the 14 years abundance is positively correlated with latitude, and in 12 of the 14 years negatively correlated with distance from the northern edge. In two of the cases (for latitude in 1985 and 1986), the relationships are significant (slope of the regression coefficient significantly different from zero at P<0.05).

The similarity of the results for different years is expected because each site has a characteristic mean index value, and so site differences tend to be retained from year to year. It seems reasonable to expect abundance of populations to decline as the edge of the range is approached, but the opposite relationship is shown. This unexpected result may be partly because habitat type and latitude are confounded; for example, chalk grassland sites, where the hedge brown is usually rather scarce, are mostly in the south. However, it seems clear that there is no tendency for the hedge brown to occur in smaller populations towards the edge of its range.

Changes in abundance

The main purpose of the butterfly monitoring scheme is to monitor changes in numbers from year to year. Fluctuations in abundance of the hedge brown have been generally similar at most sites, with peak numbers in the early 1980s, followed by a decline from 1984 to 1988 (Table 1). An increase in numbers in the warm summer of 1989 still left the collated 'all sites' index values well below the level at the start of recording in 1976. Thus, there is no indication that the overall abundance of the hedge brown has increased during the period of expansion.

Effects of weather on abundance

The synchrony of changes at different sites and in different regions (Table 1) provides evidence for a strong effect of a widespread factor on numbers of the hedge brown, and this factor is almost certainly weather. It has been shown that for many butterflies there are strong correlations between changes in numbers and weather; in several species, numbers tend to increase in warm summers. Pollard (1988) showed that 75% (as shown by the coefficient of determination) of the fluctuations in hedge brown index values from 1976 to 1986 could be explained by a model which included the previous population index (within the model) and mean summer (June, July, August) temperature (Figure 1).

When used to predict the 1987–89 index values, the model predicted the general pattern of fluctuations, but not the level of index values. Given the broad measure of temperature used in the model and the

Table 1. Synchrony of changes at different sites and in different regions. Index values for the hedge brown from 1976 to 1989; data for ten sites in the monitoring scheme for which index values could be calculated in all years. The correlation coefficient (r) is between index values for the site and collated 'all sites' index values, excluding the particular site. Significance of r indicated by $*P<0.05$, $**P<0.01$, $***P<0.001$. The generally high r values show the strong synchrony of trends and suggest the operation of a widespread factor, probably weather

Sites	Grid reference (10 km squares)	1976	77	78	79	80	81	82	83	84	85	86	87	88	89	r
Aston Rowant (Beacon Hill)	SU79	29	28	37	119	100	132	125	157	247	78	33	11	10	21	0.816***
Aston Rowant (Bald Hill)	SU79	5	23	22	101	55	46	106	40	108	105	35	7	6	8	0.629*
Castor Hanglands	TF10	264	261	164	167	142	238	540	379	808	412	339	118	18	141	0.847***
Dyfi	SN69	197	240	75	45	82	60	292	433	218	206	73	145	105	89	0.515
Kingley Vale	SU81	786	405	302	397	245	438	734	899	1 690	303	136	95	75	148	0.942***
Monks Wood	TL27	294	247	214	354	307	334	559	423	746	262	209	99	60	105	0.912***
Studland Heath	SZ08	223	158	78	80	91	142	270	435	554	333	128	151	211	157	0.584*
Walberswick	TM47	577	80	162	297	169	175	329	621	542	304	141	73	90	390	0.772**
Woodwalton Farm	TL28	323	189	233	235	235	201	194	247	262	161	158	59	91	150	0.656*
Yarner Wood	SX77	95	56	99	51	54	143	216	176	384	193	56	45	51	82	0.817***

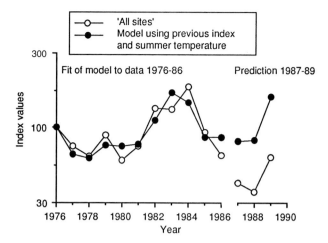

Figure 1. Hedge brown collated index values ($\log_{10}$) from all sites in the monitoring scheme from 1976 to 1986 are used to obtain the multiple regression equation $Y=1.10+0.21X+0.17T$ (Y=current index, X=previous year's index, T=mean June, July, August 'England and Wales' temperature (see text). The fit of the model to the 1976–86 data is shown; the amount of variability accounted for (r^2) is 75%. The partial regression coefficient of T is significant at $P<0.01$. The model is used to predict the 1987–89 data. The 1989 temperature data used are approximate

few years' data on which it is based, the agreement was probably as close as could be expected.

Considering the positive association between summer temperatures and hedge brown numbers, and the slight tendency for numbers to decline over the recording period, it is not surprising to find that summer temperatures during the recording period have generally been low. In fact, nine of the 14 summers in the recording period were below the 1951–80 average (Meteorological Office 1976–89). Thus, it seems unlikely that the recent expansion of range of the hedge brown is related to changes in abundance or to favourable weather.

However, although the monitoring scheme data provide no clear evidence that the expansion of range is related to weather, such a relationship cannot be completely discounted. The monitoring scheme has been in operation for only a part of the expansion period, and it is possible that the abundance of the species increased greatly during the early period of expansion and that numbers are still high relative to earlier decades. Also, as mentioned above in relation to distribution records, spread may have been in one or a few exceptional years which favoured dispersal, but the spread was detected over a longer period. If this hypothesis were true, no relationship with general abundance need be expected.

PHENOLOGY

The regular counts made through each recording season in the butterfly monitoring scheme have amassed a large body of data on the flight periods of butterflies. Brakefield (1987) previously examined these data for the meadow brown (*Maniola jurtina*) and the hedge brown, and described the flight periods in terms of mean flight data and standard

deviation about the mean date. It has been shown that the standard deviation (SD) is, in the case of the hedge brown, closely correlated with the length of the flight period (Pollard 1991).

Brakefield found that the flight period of the meadow brown was markedly shorter in the north of Britain than in the south, but that of the hedge brown was much less flexible. He suggested that the more restricted range of the hedge brown might be a result of this inflexibility; that is, it was unable to adjust its flight period to a length suitable for conditions further north.

Brakefield showed that the timing of the mean flight date was strongly correlated with temperature during the period before the emergence of the butterfly. The emergence of the hedge brown usually begins during late June or July; Brakefield showed a correlation coefficient (r) of −0.87 between flight date and mean June temperature. He also found a tendency (non-significant) for the flight period to be shorter in warm seasons. The analysis of the phenological data of the hedge brown has now been extended to examine temporal trends in mean flight date and the length (SD) of the flight period from 1976 to 1989 at the 35 sites considered above (Pollard 1991). An outline of the results will be given here.

Although Brakefield (1987) found no significant shortening of the flight period in the north, there was a non-significant tendency for such a shortening. In the longer period of data examined by Pollard, the flight period was found to be significantly shorter towards the edge of the range, although this trend was much less pronounced than in the meadow brown.

Considering all the 35 sites, there was a significant, although weak, tendency for the mean flight date to become earlier and a strong tendency for the flight period to become longer over the 1976–89 period. The change in mean date was in the order of three days. The change in flight period, based on four standard deviations (approximately 95% of the flight period) was in the order of five days. At four of the 35 individual sites, the flight period became significantly (P<0.05) longer over this period. The trend was evident when the temperature, which has a strong influence on mean flight date and a weak influence on the length of the flight period, was taken into account. During a period of expansion of range, this result provides some support for Brakefield's suggestion that there may be an association between range and flight period.

DISCUSSION

The information from recording and monitoring schemes shows that, during a period of expansion of range, there is little or no evidence for a relationship between range and abundance, but there is an association between range and flight period. There has been a trend towards an earlier flight period, but, in particular, there has been a clear lengthening of the flight period. It appears to be the first time such an association has been demonstrated, although, of course, detailed data on both factors have rarely, if ever, been studied previously.

The extension of range and lengthening of the flight period, although concurrent, may be unrelated. However, the fact that the flight period tends to be shorter close to the edge of the range does provide additional evidence that lengthening of the flight period may have been a cause of expansion of range. Brakefield (1987) suggested that the range of the hedge brown may be restricted by the inflexibility of the flight period; these results suggest that the flight period is more flexible than Brakefield thought, but strongly support his view.

The finding of a possible relationship between flight period and expansion of range leads to further questions. Do other species show similar effects? Are contractions of range accompanied, in some cases, by shortening of the flight period? What is the cause of such phenological changes? Data are available from the butterfly monitoring scheme to examine some of these questions. However, the main purpose of this account is to show how information from two long-term recording schemes can be combined to explore the relationships between distribution, population dynamics and phenology. Survey and monitoring have been shown here to be capable of drawing attention to novel features of these relationships, and suggest directions for future research.

ACKNOWLEDGEMENTS

I would like to thank the Nature Conservancy Council and the Institute of Terrestrial Ecology for their support of the butterfly monitoring scheme since its inception. I am grateful to B C Eversham for information on the spread of the hedge brown, K H Lakhani for suggesting the form of the model used in Figure 1, T J Yates for help in many ways, and the many recorders, without whom the study could not have been made.

REFERENCES

Beirne, B.P. 1955. Natural fluctuations in the abundance of British Lepidoptera. *Entomologist's Gazette*, **6**, 6–52.

Brakefield, P.M. 1987. Geographical variation in, and temperature effects on, the phenology of *Maniola jurtina* and *Pyronia tithonus* (Lepidoptera, Satyrinae) in England and Wales. *Ecological Entomology*, **12**, 139–148.

Brakefield, P.M. & Emmet, A.M. 1989. *Pyronia tithonus* (Linnaeus) – the gatekeeper or hedge brown. In: *The moths and butterflies of Great Britain and Ireland: Vol. 7, Part 1, The butterflies,* edited by A.M. Emmet & J. Heath, 267–269. Colchester: Harley Books.

Fust, H.J. 1868. On the distribution of Lepidoptera in Great Britain and Ireland. *Transactions of the Royal Entomological Society of London*, **4**, 417–517.

Heath, J., Pollard, E. & Thomas, J.A. 1984. *Atlas of butterflies in Britain and Ireland.* Harmondsworth: Viking.

Meteorological Office. 1976–89. Monthly weather reports. London: HMSO.

Pollard, E. 1979a. Population ecology and change in range of the white admiral butterfly *Ladoga camilla* L. in England. *Ecological Entomology*, **4**, 61–74.

Pollard, E. 1979b. A national scheme for monitoring the abundance of butterflies: the first three years. *Proceedings and Transactions of the British Entomological and Natural History Society*, **12**, 77–90.

Pollard, E. 1988. Temperature, rainfall and butterfly numbers. *Journal of Applied Ecology*, **25**, 819–828.

Pollard, E. 1991. Changes in the flight period of the hedge brown butterfly (*Pyronia tithonus*) during range expansion. *Journal of Animal Ecology*, **60,** 737–748.

Pollard, E., Hall, M.L. & Bibby, T.L. 1986. *Monitoring the abundance of butterflies, 1976–1985.* Peterborough: Nature Conservancy Council.

Pratt, C.R. 1986–87. A history and investigation into the fluctuations of *Polygonia c-album* L.: the comma butterfly. *Entomologist's Record and Journal of Variation*, **98**, 197–203, 244–250; **99**, 21–27, 69–80.

Legislation and policies

Managing the changes in British wildlife: the effects of nature conservation

The role of national and international wildlife legislation and of the voluntary conservation movement in the protection of species, habitats and sites

D R Langslow
Chief Executive, English Nature, Northminster House, Peterborough, Cambs PE1 1UA

INTRODUCTION

This contribution is optimistically titled; all too rarely does the opportunity to manage change occur. Almost invariably, management of the consequences of change is the only option and nature conservation is often forced to be reactive. This is a pity, but it will play that role until environmental considerations are built in as a primary consideration for managing our land and natural resources.

This paper describes some of Britain's wildlife legislation and gives examples of its effects, before turning to international wildlife legislation, which came somewhat later and has rather different implications. Some general conclusions are drawn about the effects of legislation, before examining the role that the partnership between the statutory nature conservation organisations and the voluntary conservation movement plays. The paper concludes with some comments about how the 1990s will need to be tackled by the nature conservation movement.

BRITISH WILDLIFE LEGISLATION

Some examples of wildlife legislation in the United Kingdom are listed in Table 1. The earliest conservation Act was probably the 1880 Wild Birds Protection Act. The earliest Act directly concerning mammals was the Grey Seal Protection Act 1914, although there were regulations and statutes about mammals throughout the 19th century. The most recent major Act was the Wildlife and Countryside Act 1981. In between there have been a variety of Acts with different purposes but, apart from Part III of the National Parks and Access to the Countryside Act 1949, virtually all the emphasis has been on individual species. The 1981 Act was a watershed for nature conservation in this country. Although Section 23 of the 1949 Act provided for Sites of Special Scientific Interest (SSSI), it dealt only with development proposals subject to planning permission, while the 1981 Act, for the first time, produced a legislative mechanism for conserving SSSIs threatened by agricultural and forestry activities.

The majority of British wildlife legislation has been concerned with protecting species, and primarily

Table 1. Some examples of key nature conservation legislation in the United Kingdom

Game Act 1831
Wild Birds Protection Act 1880
Grey Seal Protection Act 1914
National Parks and Access to the Countryside Act 1949
Protection of Birds Acts 1954 to 1967
Countryside Acts 1967 and 1968
Conservation of Seals Act 1970
Badgers Act 1973
Nature Conservancy Council Act 1973
Conservation of Wild Creatures and Wild Plants Act 1975
Endangered Species (Import and Export) Act 1976
Wildlife and Countryside Act 1981 (and Amendment) Act 1985
Environmental Protection Act 1990

individuals. If a species suffers persecution, and this is a major cause of mortality, then protective legislation of individuals can be extremely helpful in arresting the decline of a population. If the population of a species is small and declining, any reduction in mortality can be beneficial. The weakness is obvious; if the protected individuals are forced to try to exist in a hostile environment, then protection alone will fail. Early legislation did not take this factor into account and, although the 1949 Act brought in some habitat protection, we had to wait for some of the international Conventions of the 1970s and the development of the Wildlife and Countryside Act before this basic ecological fact was fully recognised in legislation. Even when persecution is not a major cause of mortality, legislation can help, because it gives the species a special status in people's eyes. It supports other conservation measures by emphasising the need for them and highlighting that a particular species may require active help.

Effectiveness of British legislation

The Grey Seal Protection Act of 1914 was the first wild mammal conservation Act in Britain. The grey seal (*Halichoerus grypus*) population was then believed to stand at only 500 individuals: now it is around 92 000. Legal protection seems to have played a major part in allowing the average 6% annual increase in population, and it has been accompanied by an increasingly high regard in the public mind.

Otters (*Lutra lutra*) present a rather different picture. Otters did not receive legal protection in England and Wales until 1978 and in Scotland until 1982. When the organochlorine insecticides, aldrin and dieldrin, began to be used in 1956, otters declined over large parts of Britain. There was also some evidence that populations were stressed by hunting and persecution. The bans on organochlorine insecticides achieved in the late 1960s and early 1970s gave the population a chance to recover, but the low point in the otter population in Britain was around 1979 and it is only now beginning to recover. There seems little doubt that the prevention of legal killing in the 1980s has assisted its recovery.

Birds have, of course, benefitted from wildlife legislation for many years. Indeed, it was the protection of birds from the ravages of collectors of feathers for women's hats that led to the formation of the Royal Society for the Protection of Birds (RSPB) just over 100 years ago. The RSPB is so well known that its enormous contributions, initially to bird conservation, but more recently on the much wider front of nature conservation, hardly need emphasis. Its championing of bird conservation has perhaps been the most significant factor for nature conservation in the entire conservation movement.

There has been a wide range of legislation on species and much of it was consolidated in the 1981 Wildlife and Countryside Act. Some 95 species of animals, other than birds, have some protection under the Act, ranging from restrictions on sale to complete protection. In addition, 93 species of plant are now protected against deliberate destruction. In contrast to the listing of specifically protected species of animals and plants, birds are treated comprehensively under the Act. The legislation begins with the statement that all birds are protected under the law, and Schedules 1 and 2 of the Act modify this statement. Species on Schedule 1 are protected by special penalties whereas Part II of Schedule 2 provides a list of species, commonly known as the 'pest list', which can be killed by authorised persons.

An important addition to the 1981 Act was the inclusion of a section making it an offence to destroy or damage a place or structure occupied by a protected species of animal. This addition has been particularly significant in relation to bats. The Act set up a framework which allows the staff of the Nature Conservancy Council (NCC) to talk with the owners of bat roosts and to try to allay their concerns. We try to persuade people to let the bats stay in their roofs, allow their young to become independent, and later leave the roost of their own accord. One of the great pleasures to many of our staff is that large numbers of home-owners, having learnt something about the bats, decide that they want to retain them. By record-ing all the enquiries to us, we can estimate the number of bats which benefit from this protection. In 1982–83 this figure amounted to 15 000 bats left undisturbed, with a further 13 000 allowed to disperse naturally with their young at the end of the breeding season. In addition, many more bats were saved from the toxic effects of remedial timber treatment or from disturbance caused by badly timed building or repair operations. Incidentally, the recording scheme has also collected much useful biological information about bats.

Summary of effects of British legislation

The benefits of species protection may be summarised by stating that it has:

- halted the decline of some species and resulted in population increases in others;
- raised the profile of many species in the public eye and helped to highlight their problems;
- led to increased effort in scientific research and survey;
- supported publicity campaigns not only to increase awareness of threatened species, but also to help enlarge the membership of specialist societies;
- had indirect effects on habitat conservation, particularly where protected species occur on sites which are not subject to statutory protection. Many populations of great crested newts (*Triturus cristatus*) have benefitted; the abundance and overall distribution of this species has undoubtedly been helped by their being placed on the relevant Schedule.

NATURE CONSERVATION LEGISLATION FOR HABITATS

Strong and specific protection of habitats is a relatively recent phenomenon in the legislation. The Wildlife and Countryside Act was the first substantial Act whereby the NCC, the Government's statutory nature conservation adviser, was given powers which enabled it to conserve SSSIs effectively. The NCC had duties under the 1949 Act to notify areas of special interest to the relevant planning authorities because, when the 1949 Act was passed, it was widely believed that the threats to most special sites came from developments requiring planning permission. Arguably we were slow to react in this country to the changes occurring in agriculture in particular, and to other land use changes more generally. The pressure on semi-natural habitats had become acute by the 1970s, and the 1981 Act sought to provide a mechanism to arrest their decline.

As an example of recorded habitat change, Dr N W Moore's extensive studies on the heathlands of southern England are outstanding. They provided the first really lucid account of the changes occurring on heathland, recording not only their reduction, but also their fragmentation (Moore 1962). His studies from the 1950s have provided a key baseline against which further change has been measured. Our conservation problems on the southern heathlands remain to this day (Webb 1986). The problems may have eased on the sites which have become nature reserves, but the pressure is relentless on the areas outside SSSIs, and even SSSIs have been far from immune from damage over the last decade.

The habitat provisions of the Wildlife and Countryside Act 1981 should be classified as successful overall. They have undoubtedly improved the conservation future of SSSIs, although they have not solved all the problems. Effectively the Act made no change to their status under the planning legislation, but it had a marked effect on how agriculture, forestry and other land uses could affect them. There is evidence that the rate of damage to SSSIs has declined under the 1981 Act, although the quality of information available is such that firm conclusions are difficult to draw. In recent years the NCC has carefully recorded all damage to SSSIs; while it remains true that no SSSI has been completely destroyed once notified under the 1981 Act, a number have been seriously damaged and there has been a change in the pattern of damage. This change can be illustrated by contrasting the recorded pattern of damage in 1980 with that in 1988–89. There has been a decline in the proportion of sites damaged by agricultural activities, but there has been a striking increase in the proportion of SSSIs damaged after the granting of planning permission. Thus, the damage is the result of an entirely lawful activity. This evidence suggests that the importance of SSSIs as the core of the nation's natural heritage is not being accorded sufficient weight in planning decisions.

One of the changes needed in the future is to make environmental impact assessment more effective and to require developers to submit alternative locations for their development rather than simply doing a damage limitation exercise on a chosen site.

One further comment before I turn to international issues. Arguably one of the weaknesses of the 1981 Act is that there is no provision for the positive management of SSSIs. So much of our landscape is man-modified that human intervention is often needed to maintain the special interest. While the provisions of the Act tackle the negative effects – the 'potentially damaging operations' – they do not provide for the positive requirements; this is another issue for the 1990s.

INTERNATIONAL NATURE CONSERVATION LEGISLATION

International legislation has played a significant role in the United Kingdom over the last 20 years. With the coming of the Single European Act in 1992, there is no doubt that the influence of the 'European Club' will become even stronger, so that most of our environmental activities will be within the framework agreed by the European Parliament and the Council of Ministers in the European Community (EC).

Table 2. International wildlife legislation. The first date is when the Convention was concluded and the second when the UK ratified the Convention or Agreement

Conventions

The Convention on wetlands of international importance especially as waterfowl habitat (Ramsar Convention) (1971; 1976)

The Convention concerning the protection of the world cultural and natural heritage (World Heritage Convention) (1972; 1984)

The Convention on international trade in endangered species of wild fauna and flora (CITES) (1973; 1976)

The Convention on the conservation of European wildlife and natural habitats (Berne Convention) (1979; 1982)

The Convention on the conservation of migratory species of wild animals (Bonn Convention) (1979; 1985)

European Communities Regulations and Directives

EC Regulation on the Implementation of CITES

Directive on the Conservation of Wild Birds (79/409/EEC of 2nd April 1979)

A list of the most relevant international legislation is given in Table 2. A striking feature is that so much was introduced in the 1970s. All the five major Conventions and the EC Directive on Birds were established in the 1970s, even if the UK did not ratify all of them immediately. The 1980s have therefore been a period for implementing these Conventions. Interestingly, while a number of these Conventions concentrate on species, most include some habitat provisions, and the Ramsar Convention on Wetlands of International Importance is perhaps the most far-sighted of them all. Not only does it deal primarily with habitats, but it also embraces the concept of wise use. It is remarkable that a Convention drawn up in 1971, and whose origins were largely species-orientated, should contain such far-sighted provisions. They have become increasingly important to the Convention during the 1980s. At the 1990 meeting of the Parties to the Convention, the wise use concept promises to be the dominating theme.

International wildlife legislation is sometimes criticised because it is difficult to enforce. This interpretation is too limited and overlooks some of the huge benefits provided by it. There are five features of international legislation which are of particular value.

1. It is a spur to national action. There is little doubt that knowing someone is watching you encourages good behaviour. Most Conventions require a nation to report regularly to an international forum and to be open to comment and criticism by others. Under EC legislation with an enforcement provision, this spur can be even stronger. Such a mechanism has been initiated in one or two notable cases in the United Kingdom, such as at Duich Moss on Islay.

2. It promotes the recognition of the value of wildlife sources in each party state which are of international significance. Britain's estuaries provide a good example.

3. It encourages co-ordinated action by countries sharing the geographical range of protected species. Co-ordination is vital where migratory species are concerned, but can also be important in maintaining the range of sedentary species. The framework provided by a Convention can help overcome some of the 'ups and downs' of other international relationships and, through a well-informed secretariat, provide the necessary continuity.

4. Agreed standards and common rules are set. This is particularly so with the exploitation of species. Thus, sustainability can be introduced as the guiding principle, support and help transferred between appropriate states, and controls implemented, as for example through the Convention on international trade in endangered species (CITES). Of course, political factors do play a part and the recent shenanigans over trade in ivory illustrate this fact all too well. CITES will need to reassert the importance of the Contracting Parties' Scientific Authorities as the basis for whatever actions are taken.

5. Finally, and most powerfully in the adversarial manner in which we conduct many of our affairs in the United Kingdom, an international designation can be a powerful moral force. When giving evidence in public inquiries, I have experienced the importance which the planning inspectors place on international designations. They invariably cross-examine closely on the international legislation and its relevance, and it arises frequently as a major issue in their reports. Such moral forces should not be underestimated: heavy-handed enforcement legislation is not the only way to proceed.

International wildlife legislation has had a major effect, both in this country and in many others, in helping the conservation of wildlife. Of course, all Conventions could be much more effective, but they do set up the framework, they establish the institutions, they provide contacts between scientists and administrators and they give opportunities for persuasion, cajoling and gentle intervention. We should nurture them, and the United Kingdom should take a strong lead both in implementing the provisions domestically and in assisting the less-developed countries.

ROLE OF THE VOLUNTARY CONSERVATION ORGANISATION

As a practical force, the voluntary conservation organisations (VCOs) are an extremely powerful group which contribute to nature conservation in a number of key ways. In particular, many of them have major site safeguard programmes whereby they acquire and manage sites of significance for nature conservation. This is a major activity of the RSPB, the Wildlife Trusts and the Woodland Trust.

Survey and recording have been traditional activities of many of the voluntary groups for more than 50 years. The activities of the British Trust for Ornithology (BTO), the Botanical Society of the British Isles (BSBI) and other 'species-group' organisations are well known, and many of their contributions have been mentioned during this Conference. They have had a long and close association with the Biological Records Centre, and one looks forward to the new Co-ordinating Commission for Biological Recording making a significant contribution in developing links between the various datasets so that their usefulness to nature conservation and for other environmental protection purposes becomes even greater. Some VCOs conduct research, while publicity and education are major activities for most.

One of the most significant roles of the voluntary conservation movement in Britain is to mobilise the enthusiasm, time and expertise of the thousands of amateur naturalists for the benefit of nature conservation. The sheer scale of the voluntary conservation movement and its effective organisation, notably in groups like the RSPB, make the movement a potent political force, as well as a vigorous practical force to advance nature conservation. Government listens attentively to them.

The report of the House of Lords Select Committee on Science and Technology (1990) ('the Carver Committee') on the scientific base of the Nature Conservancy Council highlighted the importance of the VCOs to the future of nature conservation in Britain and the need for the Government to provide statutory organisations in which VCOs have confidence. The comment was made both for political and practical reasons and especially because nature conservation, like many other activities in our country, depends on an effective partnership between the voluntary movement and the official bodies, such as the NCC. The NCC's strategy document *Nature conservation in Great Britain* (Nature Conservancy Council 1984) set out these partnerships and objectives in detail, and in re-reading it before preparing this talk, I was struck by how many of its messages remain as important today as in 1984. If you haven't read it recently, or even at all, do so!

The voluntary conservation organisations have had a major effect on wildlife legislation. The RSPB led the way in persuading Parliament to pass comprehensive legislation in the Protection of Birds Act 1954 and played a major role in its revision in the 1967 Act and in the 1981 Wildlife and Countryside Act. It has continued to play a significant role in influencing the new legislation affecting the environment and the role that other legislation lays in environmental protection. The RSPB has created an extremely favourable climate of opinion about bird protection within our society and has made activities like egg-collecting and the illegal taking of birds of prey socially unacceptable. The Royal Society for Nature Conservation too has played a major role recently, notably with the 1981 Act and in much of the follow-up work. Other specialist bodies like the BTO, the Wildfowl and Wetlands Trust, the Mammal Society, the British

Herpetological Society and the British Butterfly Conservation Society and others make a significant contribution and now co-ordinate themelves through Wildlife Link. No-one should underestimate the force of the voluntary conservation movement in nature conservation in Britain or, indeed, the value that organisations like the NCC place upon it.

NATURE CONSERVATION IN THE 1990s

And now what of the 1990s? The title of this talk was 'Managing the changes' – and it is, as stated at the outset, a distinctly optimistic title. We tend to react retrospectively to the changes, and hence the most important need is for us to know what the changes are and how they are occurring. Biological recording is fundamental and, within the next few years, its importance is likely to be enhanced, its co-ordination greatly improved, and the total resource available to it increased. Concern for the environment amongst the wider public has never been greater and it is a concern which is here to stay. Thus, those who govern our society will need to be able to demonstrate that their policies and programmes are *not* having a detrimental effect on the environment and that we are repairing some of the damage inflicted over past centuries. To do that effectively we need to know the extent and rate of the changes and how to manage them.

That's all very well, you may say, all very commendable, but what about the practical implications? I believe that the Brundtland report (World Commission on Environment and Development 1987) may turn out to be the most significant document of the last decade. It is the first real attempt to analyse environmental problems worldwide and to offer solutions which are practical in the political and economic framework within which we must all exist. Serious thinking about these issues by our own Government have been fed into the White Paper which Mr Patten, the Environment Secretary, produced in autumn 1990[1]. Given his previous interests at the Overseas Development Administration, his public statements and those of the Prime Minister, and the fact that his own special adviser is Professor Pearce, I am quite certain that the principles of sustainable development and a proper evaluation of all the environmental resources will play an increasingly important role into the future. All of us will need to contribute if future changes are to be successfully managed, improvements made to the quality of our environment, and especially if the nature conservation resource is to be safeguarded and enhanced. We shall have to become more imaginative in using our knowledge and information and be prepared to argue and debate, on a solid ecological basis, what is actually needed.

There is no doubt that the international scene will become increasingly important. The Single European Act in 1992 will result in some issues within the EC being decided on qualified majority voting. This will mean that, in a number of important aspects, one or two countries will not be able to refuse the introduction of new Directives and Regulations. The effect is likely to be an increase in the amount of environmental legislation emerging from Brussels, and the UK's operational framework will be determined through the European Parliament and the Council of Ministers.

The significance of European legislation, and especially of the highly influential Birds Directive, was mentioned earlier. An EC Habitats Directive is under negotiation. It would be highly beneficial to wildlife in Europe and would, I believe, significantly help nature conservation in the United Kingdom. On the wider international stage, the Ramsar Convention came of age during the 1980s and promises to be a potent force for wetland conservation in the 1990s. The International Union for Conservation of Nature and Natural Resources (IUCN) and United Nations Educational, Scientific and Cultural Organisation (UNESCO) are actively examining the possibilities of a Biodiversity Convention, which might encompass much existing international wildlife legislation and set out a new series of principles and interactions. The aim would not be simply to replace existing legislation or indeed to fill in the gaps, but to establish a comprehensive framework within which the international community would have unambiguous duties and responsibilities towards biodiversity. The contribution of biological recording to such a Convention must be obvious.

The principles of sustainable development will be part of the key framework of a Biodiversity Convention and will increasingly form part of all wildlife legislation. Within Britain, the Environment White Paper will lead to new legislation. Reforms and development of the 1981 Act have already been identified by the Nature Conservancy Council and others, and have been the subject of Select Committee Inquiries. A number of changes are acknowledged to be needed and may come as part of new environmental legislation. Environmental audit is also likely to be a feature of the 1990s and is an important part in managing changes. Effective and well co-ordinated biological recording is absolutely fundamental to it.

The 1980s were exciting for the environment, but I am convinced that the 1990s will be more exciting. The fundamental changes that have occurred in the attitudes of people and of key politicians mean that the environment will be much higher up on the political agenda in the 1990s than it has been in the past decade. I have carefully *not* mentioned global climatic change in this paper but, unless the scientists have made serious errors in their interpretation, we can look forward to a period of unprecedented change in our environment. Managing that change will test to the limits the scientific capabilities and the political will of countries across the globe. It promises to be a fascinating challenge, and I certainly look forward to it.

[1] *This common inheritance* (Cm 1200, 1990). London: HMSO.

REFERENCES

House of Lords Select Committee on Science and Technology. 1990. *Nature Conservancy Council, Volume 1 – Report.* London: HMSO.

Moore, N. W. 1962. The heaths of Dorset and their conservation. *Journal of Ecology*, **50**, 369–391.

Nature Conservancy Council. 1984. *Nature conservation in Great Britain.* Shrewsbury: Nature Conservancy Council.

Webb, N. R. 1986. *Heathlands.* London: Collins.

World Commission on Environment and Development. 1987. *Our common future.* Oxford: Oxford University Press.

Environmental politics and policies

T Burke
Director, Green Alliance, 49 Wellington Street, London WC2E 7BN

I would like to build on the foundations of Derek Langslow's talk and look at the broad political context in which today's ideas are going to be played out. I am going to draw a distinction between policy and politics. Policy is the route map, how to get from A to B; politics is the actual journey; and there is all the difference in the world between the two. For the overwhelming majority of the last two decades we have had neither an environmental policy nor environmental politics. Today we still don't have a policy, but we do have the politics. Derek Langslow has given us a good example of one aspect of environmental policy. Legislation has developed piecemeal since as far back as the 14th century, including laws of nuisance, planning legislation, and countryside and wildlife legislation, and all have developed in isolation from each other. There has been no attempt to make any sense of these different aspects of environmental policy. One example of this appalling record is the Act in 1974 which pulled together various aspects of pollution policy. The legislation went on to the statute book and was promptly forgotten. Even today there are sections which have never been implemented. We have good reason to criticise this lack of environmental policy. There is no guidance on what local authorities and other organisations should be doing. All through the 1980s we had 'policy by pamphlet': pamphlets issued by the Department of the Environment (DoE). The Government White Paper published in autumn 1990[1], therefore, was quite a momentous step, pulling together all the environmental strands of existing legislation. It was the first-ever White Paper on the environment. For the last 20 years the DoE has been operating without any formal charter, and there has been no annual report on the state of the environment.

Environmental politics in this country began on 7 October 1988, when Mrs Thatcher delivered an unexpected speech on environmental policy to a distinguished audience of scientists. The following month oilman Bush made the environment a major issue in his presidential campaign. Then, one month later, Mr Gorbachov made a speech in which he mentioned the environment no less than 25 times. So, in a very short space of time – from October to December 1988 – three major leaders had introduced the environment on to the political agenda. Before this time no national leader of global stature had ever dedicated a speech to the environment. These three speeches were followed by a series of initiatives. *Time Life* magazine, unable to find a candidate for their Man or Woman of the Year, chose the Earth as the Planet of the Year instead, and devoted a whole issue to a review of the state of the environment. Mrs Thatcher convened a conference in London on the ozone layer – and even paid the fares of most of the delegates. The summit in Paris, of seven industrial leaders in July, was devoted to the global environment. Mrs Thatcher removed Mr Ridley and brought in Mr Patten to improve the Government's 'green image'. So there has been an exceptional transformation in just over a year in the politics of the environment.

Why did it happen then, after being ignored for so long? There was no remarkable change in the environmental agenda. There were no new issues in 1988. So why did the politics change? Probably because of underlying social forces. Public opinion did not change very much, it just rose steadily throughout the 1980s. What did change, around the middle of 1988, was that people began to translate their general preference for environmental things into specific choices in the market place. *The Green consumer guide* was published, sold half a million copies and was on the best-seller list. In a survey, 42% of adults said they had already made a consumer choice by buying 'greener' products, and in 1989 battle raged between major retailers such as Safeway, Tesco and Sainsbury's, competing to become the 'greener grocer'. A small German battery company, Varta, removed a small amount of mercury from their products, sold them as 'green', and increased their sales by several 100%. People's voting choices also reflected this new trend. The UK Green Party got 15% of the vote in the European elections. The 2.3 million people who voted for them may not have understood their policies, as most people fail to understand the policies of all political parties, but they voted green because the Green Party stood for the environment. There has also been a dramatic increase in membership of environmental organisations. Friends of the Earth's membership increased from 98 000 to 182 000 between May and October 1989. And environmental organisations, after all, are simply marketing organisations for the

[1] *This common inheritance* (Cm 1200, 1990). London: HMSO.

environment. People are therefore expressing their desire for a better environment, and MPs, in speaking out in support of the environment, are legitimising and reinforcing people's inclinations so that they then become even more concerned. This development reflects a very important change in environmental politics – the public have been told that they are right to be concerned about the environment, and it would be very hard now to reverse that position. I have dealt with these recent changes regarding 'green issues' in more detail elsewhere (Burke 1990).

However, there has been no change in public policy or performance, and there are three major implications.

1. There will be much pressure to close this performance gap, and it will require a lot of knowledge, eg about the acceptable trade-offs that can be made in the future. The resource held by the biological recording community will therefore become much more important, and you must be sure that you are not giving it away too cheaply.

2. This situation will create a need for even more knowledge. Priorities in research will shift and more resources will be directed towards fundamental research. However, there will be many more hands scrambling to get those resources, so the recording community must make sure it argues loudly for its share.

3. The role of science within the public debate will be altered. Until now the scientific community has been timid, although it has begun to change. But the debate will broaden to address much more general issues, and science must be visible within this broader context. You must be prepared to say that there is not enough investment in science education. You must be prepared to say that there is not enough investment in scientific research. Research is currently concentrated on short-term issues and is neglecting the fundamentals, particularly in the biological sciences. It is extremely important to be equipped with biological knowledge, but it won't happen if you don't shout.

REFERENCE

Burke, T. 1990. *Anno viridis* – the green revolution in Britain. *Landscape 90, – The Environmental Review*, **2**, 81–84.

Poster session

The library of the Linnean Society, the venue for the refreshments and conversazione during the conference, was taken over, for the day, with posters on topics relevant to the meeting. Although several posters related to work covered by papers given during the day, brief summaries of seven additional posters are included in the following section.

The work of the National Trust Biological Survey Team

K N A Alexander
The National Trust, Spitalgate Lane, Cirencester, Glos GL7 2DE

In 1979, the National Trust set up a Biological Survey Team to identify the biological interests of its land and to produce outline management prescriptions for their conservation. The Team is multidisciplinary, covering botanical, zoological and geological aspects.

The Trust owns a considerable amount of land – about 250 000 ha. Hence, only brief reconnaissance surveys have been carried out. At each property, the vegetation and broad habitat types are described and their boundaries mapped at 1:10 000 scale. Botanical recording is mainly of vascular plants. Zoological survey is carried out on a representative selection of the habitats, invertebrate sampling being carried out by sweep-netting, beating, hand search, and direct observation. Strong emphasis is placed on searching for species regarded as good indicators of habitat quality. Sightings and signs of vertebrates are also recorded. No attempt is made to compile comprehensive species lists. Field work is normally carried out between May and September, and is supplemented by an enquiry of existing data sources, consulting relevant organisations and individuals and through literature searches.

All available information is written up as individual property reports, with 650 written at the time of Conference. Reports are as fully integrated as possible; they are targeted at the managing agents, who have management responsibility for the properties, and also the wardens, who are in day-to-day control. They are used in the production of property management plans which integrate diverse interests, such as landscape, wildlife, archaeology, public access, agriculture and forestry. Copies are sent to the Nature Conservancy Council, the local County Trust, and the local records centre. Where inadequately covered interests have been identified, further specialist survey may be arranged, using volunteers and contract support.

The first round of surveys was completed in 1987 and written up by 1989. In that year a return cycle of visits was initiated to update the property descriptions and management advice. Of the 31 properties resurveyed – all in Cornwall, 68% had changed so much that previous management advice was no longer valid. The rough clifflands are still a major refuge for species-rich plant and animal communities, but lack of grazing and uncontrolled fires have led to dramatic changes in structure and composition of vegetation at many sites, and hence also in the associated fauna. Other clifflands are adequately grazed and in good condition. Some woodlands have been degraded, but others are much improved through appropriate management. Many old grasslands survive within farms, others have been lost.

Return visits have provided opportunities to incorporate newly available information – both new site information and also wider research which helps place individual sites into better perspective, eg national surveys of rare species, the National Vegetation Classification, and increased appreciation of mature timber communities. The continued work of the Team also allows new acquisitions to be surveyed.

The breadth of experience within the Team enables it to provide a national overview of the Trust's land and to place each individual property in perspective. The Team's work has had a major impact on knowledge and appreciation of nature conservation within the Trust and of how its land is managed.

National rare plant surveys

R Fitzgerald and L Farrell*
Nature Conservancy Council, Northminster House, Peterborough, Cambs PE1 1UA
*Address for correspondence: English Nature, Northminster House, Peterborough, Cambs PE1 1UA

Britain must be one of the best 'botanised' areas in the world, with a particularly fine network of recording field botanists and taxonomists, and a good communications system through the Botanical Society of the British Isles (BSBI) and other national and local societies. With a relatively small flora, because of the recent geological history of these islands, and a large number of experienced botanists, it is not surprising that the distribution of vascular plants is well understood. Sites for rare plants are often famous (some are a little too famous for their own good), most counties have at least a published historical *Flora*, and there are numerous other publications.

Despite this apparently promising situation, it can be surprisingly difficult to assemble accurate information on rare species. The Nature Conservancy Council (NCC) is responsible for the protection of species and communities, together with the local Wildlife Trusts, and needs immediately accessible data on rare plants for the many decisions, legal and practical, which are made by its officers, and for advising the Government on protection policies. In 1974 a pilot scheme to assemble information on rare vascular plants, in East Anglia, was instigated by Lynne Farrell with Mrs G Crompton as surveyor. During the following ten years they developed the methodology used in the present rare plant surveys, which cover the *Red Data Book* (*RDB*) species defined by Perring and Farrell (1983).

Surveys are carried out in the individual NCC regions (each region normally consists of four or five counties), and the surveyor examines the history of all *RDB* species which are, or have been, recorded in the region. Current records are consulted, and visits made to extant sites, which are recorded with maps and site photographs in such a way as to make them easy to refind. Many former records are very imprecise, such as 'near Exeter', so those generated by a rare plant survey are designed to give precise information on localities in terms of landmarks and vegetation type, and a clear visual record of how the site is placed in the landscape. Population estimates give a basis for future monitoring, and threats and site management problems are identified. Although the surveys are primarily concerned with *RDB* species, the surveyor has an opportunity to identify important local rarities and species under threat, and to bring these to the notice of NCC.

Researching a survey is a complex matter. A basis can readily be established from published *Floras* and from the Biological Records Centre, but more detail has to be gathered from herbaria, journals and correspondence. Ecological information must be assembled, if it exists, and the surveyor needs to become familiar with often subtle habitat characteristics. Species are rare for a variety of reasons, such as climate, geographical range, habitat destruction, as well as detailed geological and edaphic factors, all of which need to be assessed for each species. Reasons for increase or disappearance are indicated where possible, and the management of some sites considered. Some of this work can be done as a desk study, but much involves field work in which the surveyor needs contacts with local botanists who may have invaluable local knowledge of the species.

Sites become 'lost' for several reasons – secrecy, accident, inaccurate records, and some plants are simply not very 'popular'. Grasses sometimes suffer in this way, and members of the Chenopodiaceae. The surveyors do not always have time to do more than record known sites, but the communication network, the cross-currents of information and the general raising of consciousness on rare species which a survey engenders, often set off a chain reaction, so that rediscoveries and new finds are made in following years.

The work of the rare plant survey is like piecing together a jigsaw puzzle. Each species is its own individual puzzle, and the survey as a whole is gradually being completed, with surveys for Wales and England already undertaken and those for Scotland planned. Although it is nearly 20 years since the scheme began, and some information already needs updating, there is a comprehensive existing basis for future work, consisting of historical and field records, ecological and management information, in an accessible form. Even though the British flora is small, its position at the geographical limits of Europe involves many species at the extremes of their range, so any work on rarities becomes part of the even bigger jigsaw of Continental vegetation.

REFERENCE

Perring, F.H. & Farrell, L. 1983. *British Red Data Books: 1, Vascular plants.* 2nd ed. Lincoln: Royal Society for Nature Conservation.

Environmental and ecological applications of invertebrate distribution data from the Biological Records Centre

M L Luff,[1] S P Rushton,[1] M D Eyre[1] and G N Foster[2]
[1] *Department of Agricultural & Environmental Science, University of Newcastle upon Tyne*
[2] *Environmental Sciences Department, The West of Scotland College*

The poster described work to develop the use of invertebrate distribution data, from Biological Records Centre (BRC) schemes, for applications such as conservation evaluation, environmental impact assessment and land use models.

TAXONOMIC GROUPS USED

Ground beetles (Carabidae); terrestrial and predatory species. Many data are available from standardised pitfall catches, with associated soil measurements.

Water beetles (Dytiscidae, etc); aquatic, mainly predatory species, with extensive data, and some measures of water quality.

Woodlice (Isopoda); terrestrial decomposers, for

which the national scheme has extensive data, with associated habitat information.

ANALYSES USED

Classification: for example, the classification of assemblages of Carabidae in north-east England (Luff, Eyre & Rushton 1989), into ten habitat groups associated with soil moisture, altitude and vegetation cover.

Ordination: as used for woodlice from the 100 km grid squares 42 (SP) and 52 (TL), showing three distinct species groups, dependent on wet soil, grass/woodland, and association with man (Harding *et al.* 1991).

Constrained ordination: for example, the ordination of fenland water beetle assemblages and how these relate to environmental parameters such as vegetation, water chemistry, depth and land management (Eyre, Foster & Foster 1990).

EXAMPLES OF TYPES OF APPLICATION

Quantifying conservation criteria such as rarity and typicality Sites were assessed for rarity by scoring species lists according to local or national rarity, using BRC data. An example of ranking of fenland water beetle sites was given by Foster *et al.* (1990). Typicality is based on the distance of any site in ordination space from the mid-point of its habitat space (Eyre & Rushton 1989). Using ground beetles from 71 woodland sites in north-east England, the 99% level of typicality excluded only two aberrant sites.

Autecological studies The example was given of the response curves of carabid beetles to environmental parameters such as altitude, soils and management. *Pterostichus diligens,* an upland damp-grassland carabid, showed positive sigmoid responses to soil moisture and altitude, but a negative response to soil density (Rushton, Luff & Eyre 1991). A single axis can be constructed to represent grassland management; examples of species showing either positive, negative or optimal responses to this axis were given by Rushton, Eyre and Luff (1990).

REFERENCES

Eyre, M.D. & Rushton, S.P. 1989. Quantification of conservation criteria using invertebrates. *Journal of Applied Ecology,* **26**, 159–171.

Eyre, M.D., Foster, G.N. & Foster, A.F. 1990. Factors affecting the distribution of water beetle species assemblages in drains of eastern England. *Journal of Applied Entomology,* **109**, 217–225.

Foster, G.N., Foster, A.P., Eyre, M.D. & Bilton, D.T. 1990. Classification of water beetle assemblages in arable fenland and ranking of sites in relation to conservation value. *Freshwater Biology,* **22**, 343–354.

Harding, P.T., Rushton, S.P., Eyre, M.D. & Sutton, S.L. 1991. Multivariate analysis of British data on the distribution and ecology of terrestrial Isopoda. In: *Biology of terrestrial isopods, III,* edited by P. Juchault & J.P. Mocquard, 65–72. Poitiers: Université de Poitiers.

Luff, M.L., Eyre, M.D. & Rushton, S.P. 1989. Classification and ordination of habitats of ground beetles (Coleoptera, Carabidae) in north-east England. *Journal of Biogeography,* **16**, 121–130.

Rushton, S.P., Eyre, M.D. & Luff, M.L. 1990. The effects of management on the occurrence of some carabid species in grassland. In: *Ground beetles: their role in ecological and environmental studies,* edited by N.E. Stork, 209–216. Andover: Intercept.

Rushton, S.P., Luff, M.L. & Eyre, M.D. 1991. Habitat characteristics of grassland *Pterostichus* species (Coleoptera, Carabidae). *Ecological Entomology,* **16**, 91–104.

BSBI monitoring scheme (1987–88)

T C G Rich*
Biological Records Centre, Environmental Information Centre, Monks Wood Experimental Station, Abbots Ripton, Huntingdon, Cambs PE17 2LS

**Present address:
24 Lombardy Drive, Peterborough PE1 3TF*

The Botanical Society of the British Isles (BSBI), in collaboration with the Nature Conservancy Council, the Department of the Environment for Northern Ireland, and the Institute of Terrestrial Ecology, set up the BSBI monitoring scheme in 1986. The project had two main objectives:

– to provide information on the current status of the vascular plant flora of Britain and Ireland by means of a sample survey, and to compare the current status with that recorded for species up to 1960 in the *Atlas of the British flora;*

– to provide a baseline for future monitoring of the flora.

The sample basis of the scheme was to collect records from one ninth of the 10 km squares in Britain and Ireland, with detailed surveys being made of three selected tetrads (2×2 km squares) in each of the 10 km squares sampled.

The project had two field seasons, in 1987 and 1988, in which members of BSBI and other botanists collected nearly one million records. These data were computerised at the Biological Records Centre, where the scheme was based throughout.

A report on the scheme has been prepared and submitted to the main funding body, the Nature Conservancy Council. The data from the scheme are incorporated in the Biological Records Centre's ORACLE database. The poster illustrated some of the results and problems of the scheme.

Twenty-five years of lichen mapping in Britain and Ireland

M R D Seaward

Department of Environmental Science, University of Bradford, Bradford BD7 1DP

The British Lichen Society, founded in 1958, launched its distribution maps scheme in August 1964. The scheme has permeated many of the Society's activities, providing a focus for field meetings, promoting herbaria and literature studies, and unifying amateur and professional lichenologists (Seaward 1988).

The 1964 experimental recording card was replaced in 1968 by a more practical card listing 728 species, based on a new introductory flora, and data collection immediately accelerated. Only after the publication of a new checklist in 1980 was the card revised to include 1100 species, almost 75% of the known flora. With about 1600 species and 3850 (10 km) recording squares, data collection was a daunting task. However, at least 20% of the Society's membership has been regularly submitting records to the scheme, and many field meetings have been held in under-recorded areas. By 1989, cards existed for 93% of the 10 km squares in Britain, but for only 50% of Irish squares. The proportion of 10 km squares with over 100 recorded species rose from less than 14% in 1973 to 41% in 1989 (Hawksworth & Seaward 1991).

Literature sources and herbaria have also been investigated intensively. Published bibliographies for Ireland (Mitchell 1971) and Britain (Hawksworth & Seaward 1977) showed there to be, respectively, 422 and 2700 publications including records; publications on lichen floras continue unabated (Seaward 1988), with an estimated 870 titles published during 1975–85 alone.

By the mid-1970s, the volume of data accumulated could obviously no longer be adequately processed manually. Mainframe computer facilities at the University of Bradford were therefore employed. Initially, computer input was by punched cards, and, as field recording cards proved difficult to read, large format transfer sheets were used. This system generated raw data for the first part of the *Atlas* (Seaward & Hitch 1982). While the mainframe system still stores the expanding database, manipulation since 1983 has been by personal Tektronix 4107 linked to the mainframe, with updating validated instantaneously and accessed by species or grid reference.

New technology has transformed production methods. The Tektronix 4107 operating on Fortran 77 with GHOST 80 graphics library enables conversion of screen-displayed maps into a variety of cartographic outputs via microcomputer- or mainframe-linked printers. With these facilities, a provisional second volume of the *Atlas* was produced (Seaward 1984a, b) and a revised edition (Seaward 1985) took three weeks from data retrieval to publication, compared with two years for the manually produced first volume.

Computer technology has revolutionised the British Lichen Society's mapping scheme. If the scheme were to be initiated now, a more extensive field structure for data capture should be designed to sort by date and substratum, retaining integral site lists within the main database, which would enhance its value for conservation purposes. Current developments include the generation of species lists ranked according to their British and Irish presence/disappearance for the preparation of a *Red Data Book*.

REFERENCES

Hawksworth, D.L. & Seaward, M.R.D. 1977. *Lichenology in the British Isles 1568–1975. An historical and bibliographical survey.* Richmond: Richmond Publishing.

Hawksworth, D.L. & Seaward, M.R.D. 1991. Twenty-five years of lichen mapping in Great Britain and Ireland. In: *Mapping of lichens in Europe,* edited by V Wirth. Stuttgart: Staatliches Museum für Naturkunde.

Mitchell, M.E. 1971. *A bibliography of books, pamphlets and archives relating to Irish lichenology, 1727–1970.* Galway. Privately printed.

Seaward, M.R.D. 1984a. *Provisional atlas of the lichens of the British Isles, Vol. 2, part 1.* 1st ed. Bradford: University of Bradford.

Seaward, M.R.D. 1984b. *Provisional atlas of the lichens of the British Isles, Vol. 2, part 2.* Bradford: University of Bradford.

Seaward, M.R.D. 1985. *Provisional atlas of the lichens of the British Isles, Vol. 2, part 1.* 2nd ed. Bradford: University of Bradford.

Seaward, M.R.D. 1988. Progress in the study of the lichen flora of the British Isles. *Botanical Journal of the Linnean Society,* **96**, 81–95.

Seaward, M.R.D. & Hitch, C.J.B., eds. 1982. *Atlas of the lichens of the British Isles.* Vol. 1. Cambridge: Institute of Terrestrial Ecology.

The Rothamsted Insect Survey light trap network

I P Woiwod and A M Riley

AFRC Institute of Arable Crops Research, Rothamsted Experimental Station, Harpenden, Herts AL5 1JQ

Since 1960 a network of standard Rothamsted light traps has been established to monitor aerial populations of moths as part of a wider project to study the spatial and temporal aspects of insect population dynamics and to investigate long-term changes in insect populations (Taylor 1986). Samples are small and have been shown to have no effect on the insect populations being sampled (Taylor, French & Woiwod 1978).

The light traps are run by volunteers throughout Britain and all the larger moths (Macrolepidoptera) are identified daily. Records are entered on to the